# Hearing Conservation Programs
## *Practical Guidelines for Success*

Julia Doswell Royster and Larry H. Royster

 LEWIS PUBLISHERS

**Library of Congress Cataloging-in-Publication Data**

Royster, Julia Doswell, 1951–
   Hearing conservation programs: practical guidelines/Julia Doswell
Royster and Larry H. Royster.
      p.      cm.
   Includes bibliographical references.
   1. Deafness, Noise induced—Prevention—Standards—United
States. 2. Occupational health services—Administration—Standards—
United States. I. Royster, Larry H. (Larry Herbert) II. Title.
RC963.5.N6R69      1990
363.7'46—dc20      89-48503
ISBN 0-87371-307-9

LEWIS PUBLISHERS, INC.
121 South Main Street, Chelsea, Michigan 48118

PRINTED IN THE UNITED STATES OF AMERICA

 **Julia Doswell Royster** is a consultant working with industry and supervising industrial hearing conservation programs as president of Environmental Noise Consultants, Inc. After practicing as a certified speech-language pathologist and audiologist in hospital, rehabilitation center, and community clinic settings, she earned her PhD from North Carolina State University, where she studied applied psychology. As an adjunct assistant professor in the Division of Speech and Hearing Sciences at the University of North Carolina at Chapel Hill, she teaches industrial hearing conservation to audiology graduate students. She also directs Council for Accreditation in Occupational Hearing Conservation (CAOHC) certification courses for audiometric technicians and teaches seminars for hearing conservation personnel in occupational settings.

Dr. Royster has completed contract research projects concerning the auditory effects of noise and significant threshold shift criteria for industrial use. She is the primary data analyst for the American National Standards Institute (ANSI) S12-12 Working Group for Evaluation of Hearing Conservation Programs and has performed research using audiometric database analysis to judge hearing conservation program effectiveness. Using research grants through North Carolina State University she has investigated the problems employees experience in wearing hearing protection devices (HPDs), industrial practices in HPD utilization, psychological aspects of wearing HPDs, and the protection provided by HPDs as worn in the workplace.

Julia is a past chairman of the North Carolina Regional Chapter of the Acoustical Society of America, past member of the American Speech-Language-Hearing Association committee on hearing conservation and noise, and is active in the National Hearing Conservation Association. She is a Fellow of the Acoustical Society of America and the current chairman of its Committee on Regional Chapters. Her publications include journal articles as well as chapters in texts on industrial hearing conservation.

**Larry H. Royster**, who has been active in the hearing conservation and noise control areas for over 25 years, received his PhD degree from North Carolina State University. He was appointed the first chairman of the North Carolina Department of Labor's Occupational Safety and Health Administration (OSHA) Advisory Council, and was the author of the original North Carolina OSHA program guidelines, including the Noise Compliance Plan.

Dr. Royster has assisted industries across the United States in implementing and evaluating hearing conservation programs, and has provided in-house training for industrial personnel involved in all aspects of hearing conservation programs. He has presented over 100 seminars and given talks on topics including hearing conservation, noise and vibration control, and management of hearing conservation programs. In addition to teaching CAOHC courses, Dr. Royster has published over 100 papers on the general effects of noise and hearing conservation related areas.

A professor in the Department of Mechanical and Aerospace Engineering at North Carolina State University, Dr. Royster is responsible for teaching graduate courses and conducting research concerning the effects of noise and vibration on humans, and noise and vibration control. He has directed research projects including the determination of the effectiveness of hearing protection devices as utilized in real-world environments, the effects of noise on industrial populations, and the effectiveness of existing hearing conservation programs. One of these research projects gave him and Dr. Julia Royster the opportunity to visit over 215 industries in all regions of the United States for the purpose of collecting hearing protector utilization data and general pertinent information about existing industrial hearing conservation programs. The data derived from this study, as well as Dr. Royster's previous real-world experiences, provide the basis for this textbook.

Dr. Royster has served as chairman of the Acoustical Society of America's Technical Committee on Noise and of the American Industrial Hygiene's Noise Committee. He was responsible for organizing the North Carolina Regional Chapter of the Acoustical Society of America and is a Fellow of the Acoustical Society of America.

# Preface

This book describes the essential characteristics of effective hearing conservation programs (HCPs), the best organization of personnel to get the job done, and the specific aspects within each phase of the program that make the difference between success and failure.

A section on the key topic of audiometric database analysis is included. In the past, hearing conservationists have relied on subjective judgments, unfortunately without success, in assessing whether the HCP was successful in preventing on-the-job noise-induced hearing loss. Today, we have important new procedures that provide objective data for judging the success of the HCP in preventing occupational hearing loss. The results from these new procedures allow management to identify problems and solve them, to achieve a more cost-effective HCP.

In addition, we have included a closing chapter that discusses the issue of Workers' Compensation for occupational hearing loss. Unfortunately, some managers are more interested in avoiding financial liability for Workers' Compensation costs than in preventing hearing loss as a goal in itself. We hope the chapter on compensation will clear up misconceptions about this topic and give the reader more confidence in dealing with compensation issues.

The opinions we have expressed in this book are based on years of experience in assisting industries in implementing effective HCPs, plus field studies at industrial sites across the United States. During these experiences we had the privilege of collecting practical information from many individuals involved in industrial HCPs. In this book we have attempted to put together the wisdom of the many industrial hearing conservationists who shared their knowledge with us.

# Acknowledgments

We are sincerely grateful to the many individuals in industries across the country, particularly in North Carolina, who over the years have educated us. They have done this by saving samples of used or altered hearing protectors, showing us examples of unusual patterns of hearing change, and in other ways generously sharing their time and experience. Without their help we could not have written this book.

We also thank the E-A-R Division of Cabot Corporation for providing many of the illustrations which accompany the text.

We appreciate the helpful criticism of Elliott Berger, Manager of Acoustical Engineering at E-A-R Division, who commented on the manuscript.

# How to Use This Book

To reflect the ways in which each phase of a hearing conservation program builds upon the earlier phases, we have written this book so that each chapter builds upon preceding chapters. If hearing conservation is a new topic for you, we suggest that you review the Contents section for an overview of this book, then read it from front to back.

If you already have some knowledge of or experience with hearing conservation programs, you may be looking for information about specific areas. In this case, we suggest that you read Chapters 1 and 2, then read the Checklists at the beginning of the subsequent chapters to identify the sections which contain the information to meet your immediate needs. Then, as time permits, even the experienced hearing conservationist should read the complete manual in order to gain additional insight into how to improve or maintain the level of effectiveness of your hearing conservation program.

# We Welcome Your Feedback

As mentioned in the Acknowledgments, we owe much of our knowledge about hearing conservation to the many industrial personnel who have taught us so much over the years. Today, we are still learning from people like you who are involved in hearing conservation efforts in a variety of industrial settings. As you read this book, we would appreciate it if you could jot down your reactions to the viewpoints we have presented. If you disagree with our ideas, we would like to know. If you agree, we would like to know that, too. Please send us an informal note to share your thoughts and experiences. You can write us in care of the publisher, or at the following addresses:

Julia Royster
Environmental Noise Consultants, Inc.
P.O. Box 144
Cary, North Carolina 27512–0144

or

Larry Royster
Department of Mechanical and Aerospace Engineering
North Carolina State University
Raleigh, North Carolina 27695–7910

# Contents

# Hearing Conservation Programs
*Practical Guidelines for Success*

# Chapter 1

# Effective Hearing Conservation Programs (HCPs): Benefits and Strategies

The purpose of industrial HCPs is to prevent employees from developing noise-induced hearing loss on the job. After occupational hearing loss was recognized as a health problem, the Occupational Safety and Health Administration (OSHA) promulgated regulations[1-2] that specified minimum requirements for employers to meet. However, simply complying with the OSHA regulations does not guarantee that a program will be effective in preventing occupational hearing loss, as many unsuccessful HCPs demonstrate.

If the employer runs an ineffective HCP, there is no payback for the time and resources invested. An ineffective HCP is only an exercise in regulatory compliance. However, *it is not hard* to achieve the goal of preventing occupational hearing loss if the employer applies a few basic principles in organizing the HCP, and *it may cost no more* to have an effective program than a poor one. Our experience with industries across the country indicates that there is no significant correlation between the amount of money spent on the HCP and its effectiveness. However, if management ensures that the HCP has the desired characteristics described in this book, the program will succeed.

1

## A LOOK AT HCPs NATIONWIDE

From 1980 to 1984 the authors received a grant through North Carolina State University from the E-A-R Division of Cabot Corporation to conduct onsite interviews with HCP personnel to describe the use of hearing protection devices (HPDs) in 218 industries of all types across the United States.[3-4] In addition to the structured questionnaire results concerning HPDs, the comments of those interviewed and our own observations provided insights about common mistakes in HCP implementation and organization. Some frequent causes of HCP ineffectiveness are shown in Table 1.

**Table 1.   Common Causes of HCP Ineffectiveness**

Inadequate communication and coordination among:
  plant personnel involved in the HCP, onsite personnel and corporate headquarters.

Insufficient or erroneous information used to make HCP decisions.

No meaningful training for HPD fitters and reissuers.

Inadequate or inappropriate selection of HPDs in stock, and over-reliance on the NRR in choosing HPDs.

Failure to individually fit and train each HPD wearer.

Over-reliance on contractors to provide HCP services.

Failure to use the audiometric monitoring results to educate and motivate employees.

Failure to use audiometric data to evaluate the effectiveness of the HCP.

By avoiding pitfalls such as those shown in Table 1 and by following the tips in this book, the employer can develop an effective HCP. The policies and procedures outlined herein have proved useful in HCPs in a variety of industries around the United States and will probably apply to most production facilities. However, it is impractical to specify HCP guidelines to cover every situation, so the hearing conservationist has to use common sense in evaluating whether particular pieces of advice are workable for each plant's HCP. Local personnel can judge best how to tailor the guidelines for their facility in order to achieve the goal: prevention of occupational hearing loss.

## BENEFITS OF THE HCP TO THE EMPLOYEE

Prevention of occupational hearing loss is the primary employee benefit of the HCP, but just why is this important? Hearing loss from any cause reduces the quality of life for the affected individual. Hearing impairment interferes with normal communication, and communication is a big part of being human. For many jobs we need adequate hearing to qualify to be hired or promoted, so hearing loss decreases our employment potential. On the job we need good communication ability to give and receive instructions, use the telephone, and detect machinery sounds and warning signals. Off the job our interpersonal communication with family and friends puts much of the pleasure in our lives, and gives us a feeling of being involved with others in recreational situations and at home (Figure 1). We also need our hearing to enjoy music and the quiet sounds of nature, such as birdsongs or rustling leaves. For all these reasons and more, maintaining good hearing is an invaluable benefit.

The HCP also provides a health screening benefit for employees, since nonoccupational hearing losses and potentially treatable ear diseases are often detected through the annual audiograms.

## BENEFITS OF THE HCP TO THE EMPLOYER

The employer benefits directly by implementing an effective HCP that maintains employees' good hearing, since workers will remain more productive if their communication abilities are not impaired. Employees with good hearing are also more versatile and can be promoted to jobs where communication (especially by telephone and in meetings) is even more important. Effective HCPs can reduce accident rates and promote work efficiency, as well as reduce the stress and fatigue related to noise exposure.

The HCP is an integral aspect of the employer's overall policy toward worker health and safety practices. Employee relations

**Figure 1.** People with noise-induced hearing loss usually find it especially hard to understand the high-pitched voices of children and women. By protecting your hearing, you can prevent noise damage from affecting your family life.

are better and job turnover is lower for companies that pay attention to the working environment. Maintaining a safe and healthy workplace contributes to the company's prestige and image as a desirable employer.

Finally, in direct monetary terms, the employer may be required to spend time and resources implementing an HCP in order to comply with OSHA and state regulations. Failure to comply can result in fines and citations. If a program complies but is ineffective, these expenditures are largely wasted. Failure to provide an effective HCP also results in losses through increased liability for Workers' Compensation claims and rising insurance premiums. However, if the HCP is effective, then there is a double payback on the resource expenditure: loss prevention plus gains in productivity.

# REFERENCES

1. "Occupational Noise Exposure Standard," Occupational Safety and Health Administration, U.S. Department of Labor. U.S. Code of Federal Regulations, Title 29, Chapter XVII, Part 1910.95, *Federal Register* 39:10466 (May 29, 1971).
2. "Occupational Noise Exposure; Hearing Conservation Amendment; Final Rule," Occupational Safety and Health Administration, U.S. Department of Labor. *Federal Register* 48:9738–9785 (March 8, 1983).
3. Royster, L. H., and J. D. Royster. "Hearing Protection Utilization Survey Results Across the USA," *Journal of the Acoustical Society of America* 26 Supplement 1, p. S43 (1984).
4. Royster, J. D., and L. H. Royster. "Evaluating Hearing Conservation Programs: Organization and Effectiveness," in *Proceedings of the 1989 Industrial Hearing Conservation Conference* (Lexington, KY: Office of Engineering Services, University of Kentucky, 1989).

# Chapter 2

# Organizing Your HCP:
# Five Phases Under a Key Individual

---

**CHECKLIST FOR HCP DEVELOPMENT**

—All five HCP phases have been implemented.

—There is a key individual in charge of the HCP team.

—HPDs are potentially effective in actual use.

—HPD utilization is enforced.

—There is active communication among the HCP team members and personnel at all levels.

---

Figure 2 presents an outline of the major aspects of an HCP. The five phases of the program are shown in the first column. The second column lists characteristics that differentiate effective HCPs from unsuccessful programs. The third column indicates various company personnel who are important to HCP success. The fourth column lists influences external to the company that may also affect the HCP. Each of these aspects is discussed briefly below, and additional information is available in the original references from which this material is drawn.[1-2]

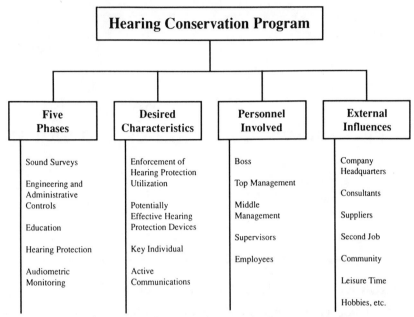

**Figure 2.** Hearing conservation program (HCP) phases, desired characteristics, personnel involved, and external influences that affect HCP function.

## FIVE RELATED PHASES

The phases of every HCP are sound exposure surveys, engineering and/or administrative noise controls, education, hearing protection, and audiometric evaluations. The relative emphasis placed on the phases may vary according to the needs of the particular production facility, but each one is essential for an effective program.

### Education

The education phase is very important because HCP personnel and employees will not actively participate in hearing conservation unless they understand its purpose and know how they will benefit directly from the program. The employer must also emphasize that compliance with the company's safety and health requirements is a *condition of employment*. Without meaningful education to motivate individual actions and consistent supervision of safety practices, the HCP will fail. Educational efforts must begin even before sound surveys and engineering controls are carried out, in order to obtain representative exposure results and to develop employee acceptance of machine modifications. Likewise, the success of the hearing protection and audiometric phases depends on teaching employees how to understand and take care of their hearing. In effective HCPs the educational phase is continuous—not just an annual presentation—as HCP personnel take daily opportunities to remind others about conserving their hearing.

## Sound Surveys

The sound survey phase involves determining the degree of hazardous noise exposure for workers so that appropriate HCP policies can be established to protect them. For example, the choices of hearing protectors available to employees may be limited to the most effective devices for departments with very high noise exposures. In addition, sound surveys can identify the dominant noise sources in each area of the plant and determine where engineering controls can significantly reduce employee exposures.

## Engineering and Administrative Noise Controls

The engineering and administrative noise controls phase attempts to reduce employees' noise exposures to nonhazardous levels. Engineering controls involve modification of the noise *source* (such as by fitting mufflers to air exhaust nozzles), the noise *path* (such as by placing sound-absorbent enclosures around equipment), or the *receiver* (such as by constructing an enclosure around the employee's workstation). Administrative noise controls include replacement of old equipment with quieter new models, establishment of equipment maintenance programs, and changes in employee work schedules to reduce noise doses by limiting exposure time. The ultimate goal is to *eliminate* employee exposures to harmful noise. However, significant *reductions* in exposure are also important because it is much easier to protect employees from lower exposures by using hearing protection devices than from higher exposures.

## Hearing Protection

The hearing protection phase of the HCP provides hearing protection devices (HPDs) for employees and training in how to wear them effectively as long as hazardous noise levels exist in the workplace. Because feasible engineering noise controls have not been developed for many types of industrial equipment, HPDs are the best current option for preventing noise-induced hearing loss in these situations.

## Audiometric Evaluations

The audiometric evaluations phase of the HCP ties together the whole program. Each exposed employee receives an annual hearing check to monitor hearing status and detect any hearing change. If the HCP is working, employees' audiometric results will not show changes associated with on-the-job noise-induced hearing damage. If suspicious hearing changes are found, the audiometric technician and the audiologist who reviews the record can counsel the employee to wear HPDs more carefully, assess whether better HPDs are needed, and motivate the individual to be more careful in protecting his hearing both on and off the job. When nonoccupational causes of hearing change are identified, such as gunfire or hobby noise exposure, or medical ear problems, they are documented and corrective actions are then implemented in order to minimize their on-the-job impact.

## CHARACTERISTICS OF EFFECTIVE HCPs

### HPDs—Effective and Enforced

The importance of HPD policies to HCP success is underscored by the first two desired characteristics of effective HCPs: *strict enforcement* of HPD utilization (actual enforcement, not just a paper policy), and the availability of HPDs that are *potentially effective* for the work environment. Potentially effective HPDs are devices which are practical and comfortable enough for employees to wear them consistently, and which provide adequate sound attenuation.

## The Key Individual

The most important strategy for making the five phases of the HCP function together effectively is to unite them under the supervision of one *key individual*. In smaller companies where one person may actually carry out all facets of the HCP, lack of coordination is not usually a problem. However, as company size increases, different types of staff become involved in the HCP: safety personnel, medical personnel, engineers, industrial hygienists, tool crib supervisors, production supervisors, etc. With personnel from varying disciplines carrying out different aspects of the program, it becomes very difficult to coordinate their efforts unless one key individual is overseeing the entire HCP. The choice of this person is critical to the success of the program. *The primary qualification for the key individual is genuine interest in the company's HCP.*

Since extra training is easy to obtain through short courses such as those approved by the Council for Accreditation in Occupational Hearing Conservation (CAOHC, shown as Resource listing #1 in the Appendix), the background of the key individual is less important than his or her enthusiasm and ability to relate to people. The key individuals we have observed know most employees by name and are equally friendly with production employees and with managers. The key individual is approachable and is always sincerely interested in comments or complaints that can help to improve the HCP. This individual does not stay in an office, running the HCP on paper by mandate, but spends some time on the production floors to interact with employees and observe how problems can be prevented or solved.

## Active Communications

The key individual maintains communication among all company personnel involved in the HCP by passing information both up and down the hierarchy. The primary HCP team members meet together regularly to update each other on the progress of the program. Once people with different tasks understand how their own parts contribute to the overall outcome of the program, they will respond to feedback about their performance by cooperating to prevent hearing loss. The key individual can achieve this active communication and cooperation if management provides him or her with the authority to make HCP decisions and the resource allocations to act on decisions once they are made.

## PERSONNEL INVOLVED IN THE HCP

The success of the HCP depends on everyone from the top boss to the most recently hired trainee; each of these types of personnel has an important role. For *top management*, the role is largely to support the HCP and enforce its policies as one facet of the company's overall health and safety program. For *middle management* and *supervisors*, the role is more direct: these staff members are part of the primary HCP team that carries out the five phases. Their duties include monitoring noise exposures, maintaining engineering controls, participating in educational efforts, fitting HPDs and reissuing HPD replacements, supervising daily HPD utilization, performing the audiometric evaluations, and giving feedback to employees about their hearing results. The role of the *employee* is to actively participate in the program and to make suggestions as to how to improve HCP operation. However, for employee participation to succeed, the HCP team must be receptive to comments and actually respond to employee input.

## EXTERNAL INFLUENCES ON THE HCP

If local HCP decisions are limited by policies mandated by corporate headquarters, the key individual may need top management's assistance in obtaining exceptions to the corporate rules in order to meet local needs. The key individual also must keep control over any services provided by outside contractors (such as sound surveys or audiograms). When contractors are used, it is more difficult to integrate their services cohesively into the overall HCP, but it is critical to do so. If in-plant personnel do not follow through by using the information provided by the contractors, then the contracted elements of the program lose effectiveness.

Finally, employees' hearing is affected by off-the-job activities such as recreational target-shooting or use of power tools in farming or woodworking. The HCP cannot save workers' hearing through *on-the-job* protection unless employees are educated to protect their ears during *off-the-job* noise exposures. The wise employer encourages employees to take HPDs home for use in nonoccupational noisy situations. Meaningful educational programs focus on the importance of good hearing for enjoyment of social and recreational activities, so the employee will appreciate the employer's concern for hearing conservation in all parts of life.

## ORGANIZATION MAKES THE DIFFERENCE

Most HCPs make at least some efforts in each of the five phases, but most programs remain ineffective because they are fragmented and incomplete. When HCP personnel lack adequate training to carry out their duties, have no direct supervision to coordinate their efforts, and are not evaluated on their performance, the HCP fails. By putting an interested and capable person in charge of all five phases, then authorizing this *key individual* to make decisions and take actions to improve the program, management will be rewarded by a cost-effective, successful HCP.

## REFERENCES

1. Royster, L. H., J. D. Royster, and E. H. Berger. "Guidelines for Developing an Effective Hearing Conservation Program," *Sound and Vibration* 16(5):22–25 (1982).
2. Royster, L. H., and J. D. Royster. "An Overview of Effective Hearing Conservation Programs," *Sound and Vibration* 19(2):20–23 (1985).

# Chapter 3

# Education and Motivation

**CHECKLIST FOR EDUCATION AND MOTIVATION**

—Team members receive education about hearing loss and hearing conservation to understand the goals and policies of the HCP.

—HCP team members receive training in how to carry out their functions (especially concerning HPD fitting and utilization).

—Employees annually attend updated educational programs which focus on why and how to protect their own hearing on and off the job.

—HCP personnel keep the program in employees' minds through informal reminders at least quarterly.

—Management backs up the HCP by personal example (wearing HPDs), policy enforcement, and participation in educational programs.

—Staff are evaluated on their HCP participation during the company's annual personnel reviews.

Education and motivation are critical in helping employees actively participate in the HCP and generating the sincere support of the program by management. Regular educational and motivational activities related to hearing conservation stimulate interest in the program and keep the importance of the HCP in mind throughout the year. Any HCP that attempts to skip over this phase of the program will find other phases failing because personnel do not understand why it is in their own best interest to cooperate in the HCP and take advantage of its benefits. More details about education and motivation are available.[1-2]

## EDUCATION PAVES THE WAY FOR OTHER PHASES

When sound surveys are planned to determine whether an HCP needs to be established, limited educational efforts must come first to notify supervisors and employees about the purpose of the sound surveys and to explain that their assistance is important for obtaining accurate sound measurements. Employees will be much more cooperative if they are informed in advance what will be happening and why.

If a noise problem is identified during sound surveys, then a more formal educational program needs to be given before the noise control, hearing protection, and audiometric evaluation phases of the HCP are initiated. This start-up program would present the results of the sound surveys, explain the risks of noise-induced hearing loss, and introduce the HCP policies established for various areas or departments of the plant.

## MAKING EDUCATION A PRIORITY

Management must emphasize the importance of the educational phase by scheduling regular training sessions and requiring attendance. Education sessions should be held not only for employees who are regularly over-exposed to noise, but also employees who occasionally enter the HPD-required areas of the plant and for the supervisors and managers responsible for production areas with hazardous noise. When a company's HCP is being introduced, a manager should participate in each educational session to outline company policies and demonstrate the company's commitment to the HCP. In an established HCP, a manager should participate in educational sessions to the degree practical to reinforce the company's priority on the HCP.

## DEVELOPING ADEQUATE PERSONNEL RESOURCES

Management must ensure that the primary HCP team members (the key individual, audiometric technicians, HPD fitters and issuers, and supervisors) have received sufficient education about hearing conservation so that they will be qualified and comfortable in carrying out their HCP responsibilities as well as in leading employee training sessions and answering employees' questions. The success of the program depends on this team.

## ORGANIZING SESSIONS FOR BEST RESULTS

The educational sessions are best structured in small groups consisting of the presenters plus the supervisor and employees in a production unit (Figure 3). These individuals will typically have common noise exposures, will fall under a common HPD policy, and will feel comfortable enough with each other to ask questions freely. Management must also ensure that employee questions and concerns raised during educational sessions receive thoughtful and prompt follow-up.

Separate sessions should be held for supervisors of noisy departments and their managers, so that these groups can discuss company policy concerns prior to meeting with employees. These groups need more detailed information to prepare them to answer questions which employees may ask them later, either during subsequent training sessions or when back in their departments.

**Figure 3.** The annual educational program covers practical aspects of all five phases of the company's HCP and gives employees a chance to ask questions.

## A RELEVANT APPROACH TO EDUCATION

To hold employees' interest, the personnel selected to make the main presentations in formal educational programs must be individuals who project genuine concern for employees' welfare. The program content must be updated every year, not just repeated.

KEEP IT . . . SHORT

SIMPLE

MEANINGFUL

MOTIVATING

Educators should limit the content to a short, simple presentation of the most relevant facts for employees. The focus should be on real-life reasons why it is to the advantage of employees to protect their hearing in order to maintain their quality of life. We need good hearing to preserve speech understanding ability in both work settings and social environments, to enjoy music, and to perceive auditory warnings and signals such as car engine noises that indicate malfunctions.

Information to help employees understand how their hearing thresholds compare to expected age-effect hearing loss will increase the motivational benefit of the audiometric phase. Once employees are familiar with their audiogram results and know the reasons why they need to preserve their hearing, the remainder of the program can focus on *how* to protect their hearing on and off the job. Employees can reduce their risk of hearing damage at work through effective use of HPDs and engineering noise controls, plus administrative controls, including good general maintenance of production equipment. The educational program should stress how the HCP benefits employees by protecting their hearing at work and detecting hearing changes which result from medical conditions and non-occupational noise exposures. A sample educational outline is shown as Table 2.

**Table 2.  Suggested Educational Program Content**

How noise damages our hearing.

Consequences of hearing loss in everyday life:
poor speech understanding,
social isolation from friends and family,
interference with work and leisure activities.

Noise exposures that are hazardous:
off-the-job (gunfire, power tools, etc.),
on-the-job (sound survey results for plant).

Engineering controls implemented or planned.

HPD choices for the employees' department:
how to use them correctly,
how to care for and replace them,
how to solve common HPD problems or complaints.

Audiometric evaluations—purpose and procedures:
Understand your own audiogram results.
Hearing changes may mean inadequate protection.
Nonoccupational hearing loss may be detected.

Ways to protect your hearing on and off the job:
Wear HPDs correctly and consistently.
Avoid unnecessary noise exposures.
Use engineering noise controls.

The company's HCP policies:
Management expresses the importance of the HCP.
HCP participation is a condition of employment.

Questions and answers.

Final motivation:
The HCP is a benefit for employees.
Participation is to employee's own advantage.

## TAILORING THE PRESENTATION TO THE AUDIENCE

Presenters need to make the educational program's content specific to the particular group of employees attending with their foreman. The program should cover the employees' specific noise exposures, the HPD options available to them, and the engineering and administrative noise controls in place or planned for their department. For the separate sessions for supervisors and managers, greater details and different emphases are appropriate to address the concerns of these personnel, such as a progress report on the status of the HCP, a review of the company's legal obligations for regulatory compliance, comparisons of audiometric and HPD utilization indicators by department, and answers for questions which employees may ask. Films and pamphlets should be used only as supplementary reinforcements for live presentations, never as the whole program. The verbal presentations and audiovisual aids should be changed each year, not repeated until the audience tunes out.

## TAKING ADVANTAGE OF MOTIVATIONAL OPPORTUNITIES

Aside from formal educational presentations, HCP personnel should use every chance to remind employees and supervisors of the importance of the HCP and their active participation in it, especially concerning hearing protection. The greatest opportunity to influence employees occurs at audiogram time, when the current hearing results can be compared to past results, and the fit and condition of HPDs can be checked. Praise for employees with stable hearing and cautions for those with threshold shifts can be effective if the comments come from a sincere individual. The personnel who really make HCPs come alive don't wait for this once-a-year chance to interact with employees; they tour the plant floor making comments and talk to workers in the halls and cafeteria. HCP personnel need to emphasize hearing conservation as an *ongoing* effort through safety meetings, rewards for departments with excellent HCP performance, bulletin board posters, articles in the company paper, and daily interactions with employees. The goal is to make the HCP a continuing emphasis that is a part of the company climate.

## QUESTIONS AND SUGGESTIONS

Employees need time during educational sessions, safety meetings, and during their daily work to voice their concerns or questions, inform HCP personnel when certain HPDs or engineering controls are not practical, and suggest alternatives which would be more workable for their departments. If HCP personnel do not provide adequate follow-up or consideration, employees need to be able, and encouraged, to go up the management line until their concerns are addressed.

## REFERENCES

1. Royster, L. H., and J. D. Royster. "Education and Motivation," in *Noise & Hearing Conservation Manual*, 4th ed., E. H. Berger, W. D. Ward, J.C. Morrill, and L. H. Royster, Eds. (Akron, OH: American Industrial Hygiene Association, 1986).
2. Berger, E. H., J. D. Royster, L. H. Royster, and D. Brus. "An Earful of Sound Advice About Hearing Protection," (Indianapolis, IN: E-A-R Division of Cabot Corporation, 1988).

# Chapter 4

# Sound Surveys

---

**CHECKLIST FOR SOUND SURVEYS**

—Representative TWAs have been determined for all noise-exposed job classifications.

—A noise map of the plant has been posted to show:
(1) areas where employees are included in the HCP
(2) areas where HPD utilization is required.

—Employees have been told the typical noise exposures for their departments during educational sessions.

—HCP team members and department supervisors have summaries of sound survey results.

—Employee TWAs are listed on their individual audiometric records.

—A report of the noise survey findings is available for review.

---

The noise measurement data obtained through sound surveys are needed to determine the degree of exposure hazard and to make decisions about how to protect employees. Different instruments and measurement methods may be used depending on the type of survey being conducted. This chapter will provide an overview, but a detailed "cookbook" approach is given elsewhere.[1]

## TYPES OF SURVEYS AND INSTRUMENTATION

In *basic sound surveys* a sound level meter is used to identify work areas which clearly *do not* have a noise problem and areas which *do* have potentially hazardous noise environments. The basic sound survey determines the departments where employees may need to be included in the HCP due to their daily noise exposures (a combination of the noise levels with their corresponding durations). In *detailed sound surveys* a sound level meter and stopwatch and/or a personal noise dosimeter are used to estimate the worker's daily noise dose and equivalent OSHA time-weighted average (TWA) noise exposure. In *engineering sound surveys* sound level meters, octave-band analyzers, recorders, and other instruments may be used to measure the noise levels produced by machinery in various modes of operation in order to assess the potential for applying engineering controls.

Surveys should be conducted on a recurrent basis—annually, or more often if it is suspected that the employees' TWAs may have changed significantly. Therefore, it is often cost-effective to purchase the instrumentation and send an onsite staff member for training in how to perform sound level and exposure measurements (Figure 4). With in-house expertise, the company can check sound levels whenever production machinery is changed without bringing in an external consultant. In addition, company personnel can evaluate simple noise control options without having to hire a consultant.

## WAYS SOUND SURVEY RESULTS ARE USED

The results of sound surveys are needed for many reasons:

- to designate those areas of the plant where hazardous noise levels exist,

**Figure 4.** A noise dosimeter takes into account the variations in sound levels over time and predicts the employee's daily noise dose, which can be converted into the OSHA TWA for the worker.

- to identify the employees to be included in the HCP,

- to classify employee noise exposures in order to define HPD policies and prioritize areas for noise control efforts,

- to determine whether noise levels present a safety hazard in terms of interference with speech communication and warning signal detection,

- to evaluate noise sources for noise control purposes, and

- to document noise levels and employee exposures for legal purposes such as Workers' Compensation.

## KEEP SURVEYS IN PERSPECTIVE

It is important to define the goals of the sound surveys and limit their scope to obtaining the information needed to guide decisions. It is not necessary to perform extremely detailed surveys in order to decide how best to protect employees; in many instances adequate data can be obtained with only a sound level meter and a stopwatch. The time and money devoted to exposure monitoring should be just sufficient to make appropriate HCP decisions, while the bulk of resource allocations should go to the phases of the program which actually *provide protection* for employees (education, noise controls, hearing protection, and audiometric evaluations). A suggested guideline for classifying HCP policies according to the established TWAs is shown as Table 3.

Table 3.   Suggested TWA Ranges for Classifying Plant Areas and Corresponding HCP Policies for Area Employees

| TWA, dB(A) | Workers Included in the HCP | HPD Utilization | HPD Selection Options |
|---|---|---|---|
| 84 or below | no | voluntary | free choice |
| 85–89 | yes | *optional | free choice |
| 90–94 | yes | required | free choice |
| 95–99 | yes | required | limited choice |
| 100 or above | yes | required | very limited choice |

*HPD utilization will be required:
1. for any individual who shows a significant hearing change,
2. for all employees if audiometric data base analysis results or group hearing trends indicate inadequate protection.

The sound survey should result in a noise map of the production facility. A noise map is a floor plan with areas of the plant designated according to whether area workers are included in the HCP, and according to which HPD utilization policy applies to the area (either voluntary or required use, with either free or restricted HPD choice, as detailed in Table 3). Estimates of daily noise exposures for representative employees in various jobs are also needed, particularly for jobs in which workers are exposed to noise which varies in level (Figure 5).

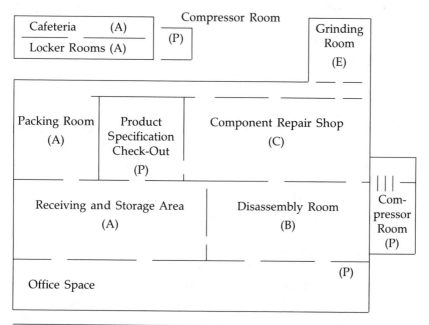

| Area Classification | TWA Range, dB(A) | HPD Utilization[1] | HPD Selection Options |
|---|---|---|---|
| A | Less Than 85 | Not Required | |
| B | 85–89 | Not Required | No Restrictions |
| C | 90–94 | Required | No Restrictions |
| D | 95–99 | Required | Limited Choices |
| E | 100 or above | Required | Very Limited Choice |
| P | | Required* | * |

[1] When an employee exhibits a significant threshold shift, then that worker's HPD options may be specified when the worker works in areas where the TWA is 85 dB(A) or higher.
*As posted.

**Figure 5.** Area classifications for different TWA ranges and corresponding HPD utilization requirements and restrictions, with floor map of one sample production facility showing the classifications applied.

## PLANNING AND COORDINATING
## WITH PRODUCTION PERSONNEL

HCP personnel must plan the sound surveys to obtain information needed to answer relevant questions about protecting employees. If external contractors actually perform the surveys, the key individual must still define the information desired and familiarize the contractors with the environment, employee work schedules, and production variations. Without overseeing the contractors, the key individual may not obtain the desired information.

The sound surveyor must coordinate scheduling with production personnel to capture representative production cycles. Supervisors can predict when noise levels will be higher or lower, when certain pieces of equipment will be in operation, and when down-time for repairs or maintenance is scheduled. Because production schedules and machinery function do not always follow predictions, the surveyors must be flexible and return as needed to obtain the desired data for all typical work activities. By coordinating with supervisors to minimize interference with production, the surveyors will enhance supervisors' cooperation and willingness to share information.

Both employees' knowledge and their cooperation are needed to obtain valid survey results. Therefore, the surveyors must also establish rapport with workers to benefit from their familiarity with the production environment and machinery. Experienced operators can often identify dominant sound sources, predict time periods of relatively higher or lower sound levels, and describe the effects of different operation modes on sound levels. By explaining the purpose of the survey to workers and soliciting their help in planning the measurements to be made, surveyors can avoid errors and oversights as well as reduce the probability of resentful or suspicious workers sabotaging the results or damaging the instrumentation.


## DATA COLLECTION

In making measurements and documenting results, the surveyors must consistently follow accepted practices for instrument selection and calibration, measurement techniques, sampling strategy, methodology description and documentation, and data recording. Detailed guidelines are available in Reference 1 and in standards published by ANSI, the American National Standards Institute (see Resources listing #2).

During data collection, the surveyor must record in detail the measurement locations and times and the sound survey procedures followed. A good rule of thumb is to make the survey description detailed enough that another person could follow it to replicate the results (provided the noise environment has not changed). It is very useful to record C-weighted as well as A-weighted sound pressure levels for purposes of estimating HPD adequacy and for engineering controls considerations.[1-3]

## EMPLOYEE PARTICIPATION IS ESSENTIAL

Employees can aid the sound surveyors in obtaining representative results by sharing their knowledge about the production environment, the machinery in operation, and worker tasks. Employee assistance is critical in operating machinery for detailed engineering sound surveys intended to evaluate sound sources within a unit of machinery. Employees also need to continue their normal activities when asked to wear dosimeters for individual worker exposure monitoring, so that the results will be representative.

Employees should be asked to notify HCP personnel when the sound environment creates a possible safety hazard due to communication difficulty, or when changes in sound levels call for a re-survey. Sound levels may increase significantly when equipment begins to wear, and changes in equipment placement or processes may have unintended effects on sound levels. When employees notice such changes, they need to inform the sound surveyors that a re-survey is needed to evaluate the sound levels and corresponding employee exposures.

## REPORT PREPARATION

The report written after completing the sound survey must present the results clearly. The writer should state the survey objectives and present data relevant to these objectives. Because few report users will need or read the full details of the survey, it is critical to include a concise abstract or administrative review section. A slightly longer summary should be included for the primary HCP team members.

The body of the report must summarize the calibration and measurement procedures to support the validity of the conclusions, and detailed documentation must be kept with the report to substantiate the procedures if they are ever questioned. Although the report will contain neatly organized tables of data, the *original* data recording sheets and instrument calibration sheets must be preserved for potential legal purposes. Keep in mind that all sound survey reports may be used as legal documents if the company ever becomes involved in a Workers' Compensation claim or other liability suit related to the noise environment. These legal actions are increasingly common.

## COMMUNICATING EXPOSURE RESULTS

The written abstract of sound survey results should be given to managers and department supervisors, and the longer summary should be given to the HCP team members. The updated noise map of the plant should be explained to employees during their educational programs and posted for employees to refer to. In areas where hearing protectors are required for all who enter, warning signs should be clearly posted. Employee TWA estimates must be transcribed onto the audiometric records for individual employees to aid the audiogram reviewer in interpreting whether hearing trends may be related to on-the-job noise exposure.

## REFERENCES

1. Royster, L. H., E. H. Berger, and J. D. Royster. "Noise Surveys and Data Analysis," in *Noise & Hearing Conservation Manual*, 4th ed., E. H. Berger, W. D. Ward, J. C. Morrill, and L. H. Royster, Eds. (Akron, OH: American Industrial Hygiene Association, 1986).

2. Earshen, J. J. "Sound Measurement: Instrumentation and Noise Descriptors," in *Noise & Hearing Conservation Manual*, 4th ed., E. H. Berger, W. D. Ward, J. C. Morrill, and L. H. Royster, Eds. (Akron, OH: American Industrial Hygiene Association, 1986).
3. Royster, L. H., and J. D. Royster. "Hearing Protection Devices," in *Hearing Conservation in Industry*, A. S. Feldman and C.T. Grimes, Eds. (Baltimore, MD: Williams & Wilkins, 1985).

# Chapter 5

# Engineering and Administrative Noise Controls

---

**CHECKLIST FOR NOISE CONTROLS**

—Engineering noise control survey has been completed and a report prepared.

—Dominant production noise sources are identified.

—Contributing equipment noise sources are identified.

—Equipment noise purchase specifications exist.

—Noise control maintenance program exists.

—HCP education phase includes engineering controls.

—New facility planning includes noise control.

—Solution of simple noise problems has been documented.

---

## ENGINEERING CONTROLS

Clearly, noise problems should be controlled if the noise sources that overexpose employees (such as production equipment, HVAC fans, or air compressors) can be quieted so that their contribution to employees' daily TWAs is no longer important, and if it costs less to control the sources than not to control them. However, real-world situations are seldom clear-cut. Making management and engineering decisions concerning the anticipated effectiveness and cost of noise control options is often a challenge to all parties involved (managers, equipment manufacturers, OSHA personnel, and consultants).

For example, consider a situation where a long-term financial benefit would result (increased production and lower cost per unit produced) if known noise control solutions for a piece of production equipment were installed. In addition, it is also known that installing the noise controls would lower employees' TWAs by at least 5 dB(A). However, the capital to carry out the modifications is not available. Therefore, in this situation the monetary constraints would delay known noise control options in favor of the utilization of hearing protection until the company is financially able to make the necessary equipment modifications. However, the higher costs of failing to make the changes as soon as practical would reduce the company's ability to compete in the free market.

What is management's responsibility with respect to the engineering and/or administrative noise control phase of the HCP? Plant management has the responsibility of identifying the dominant noise sources in all production areas and then determining if practical noise control options are available at a justifiable cost. It is not adequate for management to say, without conducting an engineering noise control survey, that production equipment A is the noise problem and no solution exists. It might be that common, easily controllable parts of the machinery (such as air exhaust or an improperly installed hydraulic valve), are the only major noise sources creating the overexposure for the operators of the equipment. Identifying the dominant noise sources in a production area is typically easy to do.

### Identifying Dominant Noise Sources

During the basic or detailed sound survey the surveyor should have identified the obvious dominant noise sources in the room.

This information is now used as the starting point for the engineering noise survey that will determine the contribution of each dominant source to employees' noise exposures.

Using a sound level meter, the surveyor can measure the sound levels at the employee's workstation as individual pieces of equipment in the production area are run separately, to determine their relative contributions. In some circumstances it is not practical to run individual pieces of equipment due to continuous operation or equipment interdependence. In such cases it will be necessary to conduct the survey during the yearly maintenance period or to make measurements at times when, due to equipment failure or temporary equipment shutdowns, one or more of the production units is not operating. Once the noise levels at the employees' workstations are known with different units in operation, then the effective contribution of each piece of equipment can be easily determined.[1]

**Contributing Equipment Noise Sources**

Once the dominant noise sources have been defined, the next step is to determine the significant contributing noise source(s) within each dominant piece of equipment. Typically this involves a team consisting of equipment operators, a mechanic, and a sound surveyor. As far as the machinery design will allow, individual components of the equipment are operated and the noise levels at the employee's workstation recorded for each operating condition. The ideal situation occurs when one component of the equipment is found to be the only significant sound source, and an inexpensive noise control fix is readily available.

**One Example**

In a production room the measured TWA of stamping machine operators was 93 dB(A), and TWAs of 88 dB(A) were measured for employees who were packing the product in another area of the room. (Note that the levels of the noise in the room were constant and roughly equal to the predicted OSHA TWA.) For employees working in both areas, noise exposures were below 80 dB(A) when the stamping machines were not running but the HVAC system was in operation. Therefore, it was concluded that the dominant noise sources in this production room were the stamping machines.

The next step was to operate one stamping machine while the remaining machines were not running. Due to production constraints, the sound testing had to be carried out between second and first shifts. The sound survey team consisted of the company's chief mechanic and the sound surveyor.

The stamping machine could be run without any product being stamped, and the air supply used to assist in the removal of the product could also be turned off. Therefore, measurements were made in the following sequence of operating conditions:

1. with only the stamping unit on (air supply turned off and no product),
2. with the stamping unit and air supply on but no product, and
3. with the stamping unit and air supply on and with product running through.

For condition 1, the measured sound level at the employee's workstation was less than 80 dB(A). When the air was turned on (condition 2), the sound level increased to 90 dB(A). Finally, for condition 3 the sound level increased by 0.5 dB(A) to 90.5 dB(A). Note that if the addition of the product (condition 3) had contributed as much to the measured noise level as did the air supply (condition 2), then the increase in level for condition 3 would have been approximately 3 dB(A). Because the increase was only 0.5 dB(A), it was concluded that the contribution from running product was at least 10 dB less than the contribution of the basic machine components plus the air noise.

After identifying the air supply system exhaust of the stamping machines as the room's major contributing noise source, management had the information to determine the feasibility of controlling the noise to reduce employees' exposures.

Since the measured sound level at the operator's workstation was 90.5 dB(A) with only one stamping machine running, and the workstation level increased to 93 dB(A) with all units in operation, then the increase of 2.5 dB(A) was assumed due to the contribution of the remaining stamping machines. In other words, noise controls needed to be applied not only to the operator's own machine, but also to all other surrounding units, in order to reduce the noise level at the employee's workstation below 85 dB(A).

In this case a solution was both economically and technically feasible. For an expenditure of less than $50 per stamping machine, management was able to completely eliminate the noise hazard. The resulting employee TWAs were below OSHA's 85 dB(A) action level, eliminating the requirement for these workers to be included in the HCP. The outcome of this engineering sound survey yielded significant long-term savings for the company. However, if management had not conducted the needed engineering noise control survey, then the information necessary to make the appropriate decision would not have been generated.

**Feasibility Considerations**

When conducting the engineering noise control surveys, it is common to find out that several pieces of production equipment are dominant noise sources and contribute roughly equally to the employee's daily TWA, and that each of these dominant noise sources includes several equally contributing component noise sources. Even when feasible solutions are known for some of the contributing sources, the existence of multiple sources often means that controlling the employee's noise exposure cannot be economically justified.

However, until the pieces of production equipment that are the dominant sound sources and their internal contributing noise sources are identified, management cannot generate the necessary technical information along with the potential effects on production and total cost information that is needed to make appropriate decisions concerning controlling the noise through engineering means.

**ADMINISTRATIVE CONTROLS**

The usual reading of the OSHA regulations limits the interpretation of administrative controls to the use of work scheduling as a means of reducing the employee's daily TWA. However, we will interpret administrative controls in a much broader and more practical sense. In addition to the modification of workers' work schedules to reduce their measured TWAs, administrative controls should also include establishing adequate equipment maintenance programs, administrative planning for noise exposures, and defining noise limits for equipment purchases and facility remodeling projects to reduce employees' TWAs (see Figure 6).

**Figure 6.** A potentially very effective administrative noise control option is the replacement of noisy equipment with quieter and more efficient new equipment.

## Controlling Employee Work Schedules

Work schedules may be modified in order to limit, or control, the employees' noise exposures. In a few special situations the use of administrative controls has not only significantly reduced employees' TWAs, but also increased productivity by sharing a very demanding task between two individuals. In one instance the operator [TWA of 89 dB(A)] and oiler [TWA less than 80 dB(A)] for a large dragline operation, who typically worked twelve hour shifts, were both retrained and allowed to change work positions every three hours. Although the oiler's salary had to be increased, the resulting increase in productivity offset the cost of higher pay for the oiler, and both employees benefited since both of their TWAs remained less than 85 dB(A).

However, caution is advised in using administrative options when they involve exposing a previously unexposed population [TWAs less than 85 dB(A)] to potentially damaging noise levels in order to reduce the TWAs for a population which is already noise-exposed. In general it is not good safety practice to increase the percentage of the workforce exposed to a known hazard.

## Maintaining an Acceptable Maintenance Program

In order to prevent the noise produced by existing equipment (newly purchased or modified by engineering controls) from increasing significantly, it is necessary to have in place a regularly scheduled equipment maintenance program. It is common to control a noise problem by engineering means only to return months later and find that the installed noise mufflers, equipment enclosures, vibration isolators, etc., have failed to maintain the controlled level of noise due to either sabotage, inappropriate equipment utilization, or inadequate maintenance. It is management's responsibility to ensure that equipment which has been noise controlled is properly serviced and utilized in order to maintain the controlled level of noise. Management typically does not realize that equipment which has been controlled for noise output must become a part of a regular noise control equipment maintenance program. Company engineers and noise control consultants have estimated that 2% to 4% of the cost of noise control will have to be spent annually in maintaining the level of noise reduction originally achieved.

It is important to require the active participation of all involved parties (equipment operators, supervisors, mechanics, etc.) in maintaining the production equipment in a satisfactory condition. This is best achieved by management directives and through the regularly scheduled company education program.

### Planning for Noise Control Purposes

Achieving a significantly quieter work environment through long-term planning (less than ten years to start of implementation) can be a very effective use of administrative noise control. It should be obvious to all by now, after over 20 years of OSHA, that in a high percentage of noisy production environments there is no quick fix available for the noise problem.

In these instances, one solution is the planned purchase of new equipment, or the remodeling of existing equipment or facilities, with sufficient guarantees to ensure that the result will be a noise-free or significantly reduced [TWA reduced by 10 dB(A) or more] noise environment. Notice that earlier a significant reduction in the noise environment was defined as 5 dB(A). However, due to the problem of estimating real-world noise environments for new production facilities, a planned reduction of 10 dB(A) should guarantee at least 5 dB(A) after equipment installation.

### Noise Limits for Equipment Purchases and/or Modifications

A second effective use of administrative controls is the establishment of *enforced* equipment noise specifications for use in purchasing new equipment or when modifying existing equipment. Notice that we emphasize "enforced" because the clear tendency in industry is for management to establish noise specifications that are regularly passed over by purchasing agents for less expensive alternatives, regardless of the cost differential.

## SOLVING NOISE PROBLEMS
## USING IN-HOUSE PERSONNEL

The preceding discussions of noise controls have been aimed primarily at defining the more obvious noise problems. Once these problems have been defined, management must decide whether to hire a consultant to achieve an engineering solution. For the simpler noise problems it is normally more cost-effective to use in-house personnel.

In order to solve minor noise problems in-house, it is necessary to:

- identify the individual who will find and implement the solution, and

- give the individual the necessary flexibility and authority.

The experiences of industrial personnel clearly show that in-house staff do not need professional training in order to solve simple noise problems. Nurses, audiometric technicians, and safety directors who have received the most elementary training in noise control concepts have succeeded in noise control efforts. Working together with the lead mechanic, they have controlled many different types of industrial noise problems, including air exhaust noise by installing commercially available mufflers; HVAC fan system noise by purchasing and installing in-line commercially available noise attenuators; and vibration-related motor noise problem by purchasing and installing appropriate commercially available mechanical isolators (see Figure 7).

For company personnel to take the attitude that noise problems can never be solved without expensive outside consulting services is simply not justified. However, experience also indicates that for the more difficult noise problems, some type of outside consultation is often necessary.

## SOURCES OF INFORMATION

For immediate information concerning possible noise control options, management has several possible sources including the resource staff of the state's Department of Labor, trade associations, insurance carriers, and extension departments at the local technical college or the state university. In addition to these resources there are articles and chapters,[1-2] plus several relatively easy-to-read textbooks[3-6] and one cost-free journal that annually publishes a summary of all the manufacturers of noise and vibration control products (see Resources listing #3).

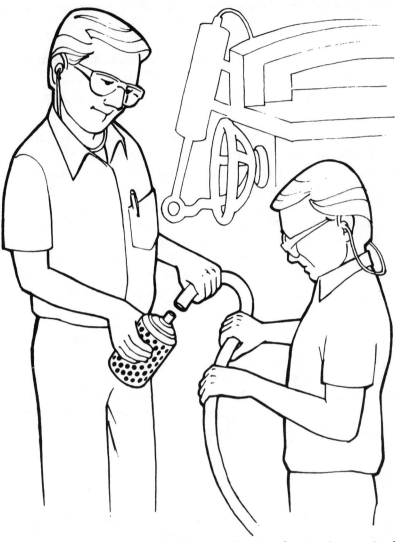

**Figure 7.** Many common noise problems, such as air exhaust noise, are simple to control with commercially available mufflers.

# REFERENCES

1. Royster, L. H., E. H. Berger, and J. D. Royster. "Noise Surveys and Data Analysis," in *Noise & Hearing Conservation Manual*, 4th ed., E. H. Berger, W. D. Ward, J. C. Morrill, and L. H. Royster, Eds. (Akron, OH: American Industrial Hygiene Association, 1986).
2. Bruce, R. D., and E. H. Toothman. "Engineering Controls," in *Noise & Hearing Conservation Manual*, 4th ed., E. H. Berger, W. D. Ward, J. C. Morrill, and L. H. Royster, Eds. (Akron, OH: American Hygiene Association, 1986).
3. Purcell, W. E. "Materials for Noise and Vibration Control," *Sound and Vibration* 14(7) 18–32 (1980).
4. Purcell, W. E. "Systems for Noise and Vibration Control," *Sound and Vibration* 14(8) 10–36 (1980).
5. Irwin, J. D., and E. R. Graf. *Industrial Noise and Vibration Control.* (Englewood Cliffs, NJ: Prentice-Hall, Inc., 1979).
6. Wilson, C. E. *Noise Control: Measurement, Analysis, and Control of Sound and Vibration.* (New York: Harper & Row Publishers, 1989).

# Chapter 6

# Hearing Protection

---

**CHECKLIST FOR HEARING PROTECTION**

—HPD utilization in required areas is strictly and consistently enforced.

—Comfort, practicality, and real-world attenuation—not the NRR—are the primary criteria for selecting which HPDs will be stocked.

—Each employee is individually fitted with HPDs and trained in their proper use and care.

—Fit is checked for all types of HPDs, including earmuffs and single-size earplugs.

—A *minimum* of two earplugs (one in multiple sizes) and one earmuff are available for selection, preferably three plugs, two muffs, one semi-aural.

—HPDs are replaced on a regular basis.

—HPD reissuers distribute only the type of HPD fitted to each employee; to change types or sizes the employee must return to the fitter.

**continued**

---

---

**CHECKLIST continued**

—Each employee's HPDs are rechecked during the audiometric evaluation for condition, fit, and correct placement.

—Employees are given HPDs to take home for use during
off-the-job noise exposures.

---

Hearing protection devices (HPDs) are the first line of defense
against noise in environments where engineering and/or administrative controls have not reduced employee exposures to
safe levels. HPDs *can prevent* significant hearing loss, but only
if their utilization is carefully implemented and supervised. Employees will not achieve adequate protection if various HPDs are
simply placed onto the tool crib shelf, with the choice of style
and size left up to the employee!

Several of the listed references offer more detailed information about hearing protection,[1,2,3] and some excellent materials
are available from Resources listing #4.

## ORGANIZING THE HPD PHASE FOR SUCCESS

HPD effectiveness cannot be achieved without the enthusias-
tic and diligent efforts of those who select, fit, issue, and reis-
sue the protectors. Management must choose capable and
interested personnel to handle the HPD phase and provide them
with the knowledge they need to do a good job. Working with
the employee in selecting optimal HPDs and training the employee to wear and care for HPDs are much more complicated
than dispensing safety glasses. Therefore, those responsible for
HPDs need detailed education from the HCP supervisor, from
attending a CAOHC course, or from materials and workshops
sometimes provided by HPD manufacturers and associations
concerned with hearing conservation.

HPDs will not protect employees unless their proper utilization is absolutely enforced as a condition of employment. Many

companies have written disciplinary rules for failure to wear HPDs, but they are never implemented. For the HCP to be effective, management must set the same priority on HPD utilization as on the use of all safety equipment, and then back it up with action: all employees (including management) either wear HPDs in required areas or they go home.

## TYPES OF HPDs

### Earplugs

The most popular HPDs are earplugs, which are inserted into the ear canal to provide a seal against the canal walls. Preformed earplugs are made of flexible, vinyl materials and often come in different sizes to fit different sizes of ear canals (Figure 8). Formable earplugs are made of materials that can be manipulated to conform to the shape of the wearer's ear canals. The best formable earplugs are made of slow-recovery foam which is compressed for insertion into the ear canal, then expands to fill the canal and seal against its walls (Figure 9).

**Figure 8.** Preformed earplugs may be available in one to five sizes to fit different ear canals.

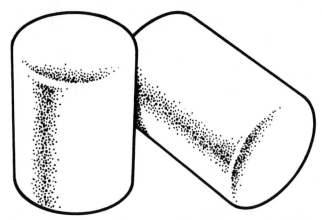

**Figure 9.** Formable foam earplugs expand to fit ear canals of different sizes and shapes.

### Earmuffs

Sometimes called circumaural HPDs, earmuffs enclose the entire external ears inside rigid cups. The inside of the muff cup is lined with acoustic foam (which must not be removed), and the perimeter of the cup is fitted with a cushion that seals against the head around the ear by the force of the headband (Figure 10). In most industrial environments earmuffs are less popular than earplugs.

### Semi-Aurals or Canal Caps

These HPDs are small stoppers which seal against the entrance to the ear canal by the force of a band which is usually worn under the chin or behind the neck (Figure 11). They generally provide less protection than earplugs or earmuffs, and they are less comfortable for long-term use. Therefore, they are most suitable for brief periods of use, not all-day wear.

**Figure 10.** Earmuffs cover the entire ear to block out noise.

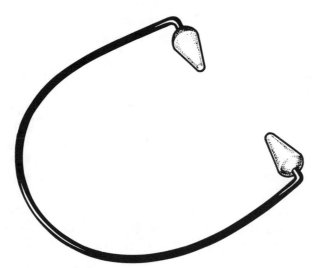

**Figure 11.** Semi-aural HPDs or canal caps are convenient for brief periods of noise exposure.

## HEARING PROTECTOR ATTENUATION

All HPDs attenuate noise by creating a barrier to reduce the air-conducted sound reaching the eardrum. The level of protection achieved depends mainly upon the degree to which the protector achieves a seal. Earplugs mainly seal against the wall of the ear canal, while semi-aurals seal against the entrance to the ear canal or its outer edge, and earmuffs seal against the skin around the external ear. In each case the amount of sound reduction achieved depends largely on the completeness of the seal—any air leaks will allow some sound to bypass the HPD.

The maximum achievable attenuation is limited by the bone-conducted sound which results when sound vibrates the skull. However, bone-conducted sound transmission is not important for industrial environments compared to the effect of whether the HPD is fitted properly and used correctly. In practice, when employees do not receive adequate protection from their HPDs, it is because they do not achieve an *adequate seal*. This may happen because the HPD *does not fit properly* (the HPD is the wrong size or design for the individual) or *is not used correctly* by the employee (due to inadequate training or carelessness) or the HPD has worn out or been abused by the wearer.

### The NRR

HPD manufacturers publish attenuation data for their products based on idealistic laboratory measurements. The Noise Reduction Rating (NRR) shown on the label of the HPD package is intended to give a single-number rating of the laboratory attenuation across a range of frequencies. The NRR is subtracted from the employee's noise exposure to indicate the maximum exposure reduction which could be attained if the employee had similar physical characteristics as the laboratory subjects and could wear the product in the same way as the laboratory subjects.

The NRR is designed to be subtracted from the C-weighted sound pressure level to give the A-weighted level under the HPD:

[Noise level, dB(C)] − [NRR] = [estimated exposure, dB(A)]

If only A-weighted exposure is known, then a correction factor of 7 dB must be subtracted from the NRR:

[Noise level, dB(A)] − [NRR − 7] = [estimated exposure, dB(A)]

The 7-dB correction factor is needed with A-weighted levels because the dB(A) value gives no indication of whether the energy in the noise environment is predominately low-frequency or high-frequency. Since HPDs provide less protection at lower frequencies, then it becomes necessary to add a safety factor unless the dB(C) value is available.

Most dosimeters predict only noise dose or equivalent TWA values based on A-weighted sound pressure levels. However, if the sound surveyor measures both dB(A) and dB(C) levels with a sound level meter (as we recommended in Chapter 4), then the average "C minus A" difference for the noise environment can be used as the correction factor in place of 7 dB. In high-speed textile spinning, for example, "C minus A" differences ranging from 2 dB to −1 dB have been measured.

## REAL-WORLD ATTENUATION

Although the NRR is a readily available number that appears as if it should allow the HPD selector to decide whether a protector is adequate, it is flawed by the laboratory test conditions used to obtain it. Real-world users do not achieve the amount of attenuation indicated by the NRR. In general, HCP personnel can count on properly trained and motivated HPD wearers receiving about 50% of the NRR value in attenuation.

The NRR cannot even be used to rank-order the real-world effectiveness of HPDs. Studies in which employees were pulled off the job to have their actual HPD attenuation measured have shown that the attenuation achieved with different HPDs gave a different rank-ordering from their NRRs. Products that are more goof-proof (earmuffs and foam earplugs) provided higher real-world attenuation than other HPDs.[3]

Because the NRR is not a realistic indicator of the attenuation which wearers achieve, HCP personnel should not use the NRR as a significant criterion for evaluating HPDs. Comfort, convenience, and compatibility with the working environment ultimately determine the protection an employee will receive from a hearing protector, since the effective attenuation of an unworn HPD is zero.

> THE BEST HPD IS THE ONE
>
> THE EMPLOYEE WILL WEAR
>
> CONSISTENTLY AND CORRECTLY!

### PURCHASING APPROPRIATE HPDs

It is essential to select and keep in stock a sufficient choice of HPDs appropriate to the work environment and the wearers' needs. Generally an adequate selection would include three types of earplugs, two styles of earmuffs, and one semi-aural device.

The choice of HPDs to be offered should be made by the HCP personnel based on characteristics of the work environment and real-world HPD performance, as well as the preferences of the employees. The purchasing department must not be allowed to overrule the HPD selection made by HCP personnel. Management needs to give HCP personnel the authority to obtain the HPDs they feel are best for the company's workforce. The factors that should be considered in choosing HPDs the company will stock are discussed in the following paragraphs.

### Real-World Attenuation

Because most employees have noise exposures below a TWA of 95 dB(A), they only need 10 dB of real-world attenuation. Therefore, they can be adequately protected by most HPDs if the devices are properly fitted and correctly worn. Higher employee TWAs require more careful HPD selection, and only the protectors with the best real-world attenuation (earmuffs or foam earplugs) are recommended for TWAs of 100 dB(A) or above.

## Comfort

Employees will not wear uncomfortable HPDs, so the goal of fitting is to find the most comfortable HPD that gives adequate protection for the environment. Because no single HPD suits all wearers, several choices should be available. The employee should be allowed to select his/her preferred HPD as long as the issuer confirms that the fit is adequate for good attenuation.

## Convenience

Employee acceptance of HPDs depends on practical factors such as:

1. ease of correct positioning, considering physical limitations of the wearer (finger size or strength, arthritis, etc.),

2. speed and ease of HPD removal and repositioning,

3. simplicity of carrying or storing during work breaks,

4. compatibility with other safety gear (masks, hard hats),

5. suitability for job tasks (crawling in tight spaces, strenuous physical activity, repetitive head movements),

6. practicality in the physical environment (for example, considering heat, dirt, chemicals, etc).

## Communication Needs

Select a HPD which will allow the employee to communicate as required in the workplace noise environment. For normal-hearing employees, this is usually not a problem since HPDs improve speech discrimination in noise above 90 dB(A) by reducing distortion in the ear from high sound levels. However, for workers with preexisting hearing loss, HPDs can make communication more difficult by reducing speech-sound information to below their hearing thresholds. Hearing-impaired employees who must receive detailed face-to-face instructions may prefer earmuffs so that they can lift up the earmuff cup to hear speech. Alternately, plugs with minimal attenuation may improve their reception of speech as well as auditory warning signals. Hearing-impaired employees may also benefit from the new HPDs presently being developed that exhibit flat attenuation across the frequency spectrum. These HPDs provide less reduction than regular HPDs of the high-frequency sounds which are affected most by hearing loss.

**Employee Input**

HCP personnel need to solicit employee opinions about various HPD products by conducting wearer evaluation trials, asking for employee feedback in safety meetings and in the break room or cafeteria, and questioning individuals when HPDs are reissued or when the annual audiogram is administered. HCP personnel should be aware of new HPDs that might be suitable additions to or replacements for the HPD choices currently carried in stock. When employees ask about products they have seen, the HPD issuer should investigate these options when practical.

## MAXIMIZING THE EFFECTIVENESS OF HPDs IN ACTUAL USE

The attention paid by the HPD issuer to the following factors will determine how much real-world protection employees receive from their HPDs.

**Correct Fit**

HPDs offer little or no protection if they do not form a seal to block out sound, so all HPDs initially should be fitted by a trained issuer. Earplugs must be fitted separately in each ear, as an individual's two ear canals may be different sizes or shapes. None of the so-called "universal-fit" plugs actually can fit every individual. Therefore, even one-size earplugs (including foam earplugs) must be checked for proper fit in each ear of each employee.

> THERE IS NO SUCH THING AS
>
> A "UNIVERSAL-FIT" HPD!

Good HCPs always stock premolded plugs in a full range of sizes to suit extra small and extra large ear canals. Before inserting an earplug into the employee's ear, the fitter must visually check the ear canal for excess wax or obvious abnormal conditions which would require fitting to be delayed until the problem is corrected. The fit of earmuffs must also be checked for each wearer to make sure that the earmuff cushion seals against the head all around the ear with adequate headband force (the wearer's head is not too narrow to permit a secure seal) and the outer ear (pinna) can fit inside the cup (the cushion must not rest on the pinna).

**Training Users**

Each HPD wearer must receive specific instructions on how to wear and care for the HPD issued. To ensure that the employee can insert or place the HPD correctly, the fitter should watch while the individual user demonstrates how to put on the HPD correctly. If the employee's initial attempt is inadequate, the fitter should reinstruct and have the worker practice in the fitter's presence until proper placement is achieved (Figure 12).

Employees must be convinced that they will not receive adequate protection unless they correctly wear and maintain their HPDs. The employee needs to know the signs of HPD deterioration which indicate that it is time to get a replacement. Each year the employee should be asked to bring his/her HPDs to the audiometric evaluation so that the fitter can inspect them for wear, reevaluate the fit, and check that the employee can still place them correctly.

**Figure 12.** Each employee needs to be shown how to insert earplugs or wear earmuffs properly to achieve a good seal and adequate protection.

### Responding to Wearer Questions and Complaints

The HPD issuers must be open to employee concerns, seeking information to answer questions and working individually with employees who find it difficult to wear HPDs with satisfactory comfort or to communicate adequately while wearing them.

### Controlling the HPD Replacement Process

HPD reissuers need to maintain strict control of replacements so that employees are reissued only the style and size of HPD indicated on the fitting record. Employees who wish to change HPD type or size must return to the fitter for refitting.

### Replacing HPDs Regularly

New HPDs should be routinely issued to each wearer on a schedule appropriate for the type of HPD being worn. An aggressive replacement procedure will prevent employees from using HPDs that have lost their effectiveness. At the same time, HCP personnel can detect altered HPDs and reeducate the employees who made alterations. Some key individuals periodically set up an HPD replacement station at the plant exit during shift changes as an extra reminder.

### Monitoring HPD Utilization

Frontline supervisors and production department heads need to perform regular checks to ensure that employees are properly wearing their HPDs. Employees who do not cooperate with the mandatory HPD utilization policy must receive reeducation and meaningful disciplinary actions, eventually culminating in dismissal for repeated offenses. At the same time, consistent HPD users should be rewarded for their performance by giving recognition to departments with good utilization records and by giving praise to individual wearers.

## MOTIVATING EMPLOYEES TO WEAR HPDs EFFECTIVELY

The employees who wear HPDs are ultimately responsible for achieving and maintaining good protection from noise for themselves through the proper use of HPDs. However, the employer must educate and motivate employees to take this active part in the HCP. The key individual and other HCP team members should strongly consider employees' input when selecting the choices of HPDs to be stocked, as well as when finding a suitable protector for an individual. Employees need clear, understandable information to help them appreciate that they will be protected from developing hearing loss *only* if they consistently and correctly wear HPDs that are fitted properly.

### Identifying Satisfactory HPDs

The employee and the fitter need to work together to select a product that will be comfortable and convenient enough for the worker to wear it consistently. The employee who has received proper education will understand that some initial discomfort is expected during the "breaking-in" period when getting used to HPDs. If the employee still encounters significant discomfort or interference with job tasks after 1–2 weeks of wearing a new HPD, then it is appropriate for the wearer to return to the fitter to request another type.

### Wearing HPDs Correctly and Consistently

Ultimately employees must accept the responsibility to reduce their noise exposure by faithfully and properly wearing HPDs both on and off the job (Figure 13). Intermittent or incorrect HPD

**Figure 13.** Shooting guns is one of the most common causes of off-the-job noise-induced hearing loss, but hearing protectors are very effective against gun noise and other hobby noises (woodworking tools, chain saws, etc.).

use will not prevent the development of noise-induced hearing loss. The department supervisors can ensure that all employees in HPD-required areas are wearing HPDs, and the HPD fitters can do spot-checks of the actual correctness of HPD fit and placement from time to time in each department. Annually, at audiogram time, each employee should be retrained in proper HPD fitting, use, and care.

## Caring For and Replacing HPDs

Employees need specific instructions as to how to wash their HPDs when needed, store them in cases or safe places to prevent damage, inspect them for signs of wear and tear, and seek replacements when deterioration occurs. Employees must be taught that achieving good protection depends on keeping the HPDs in good condition. Worn-out or intentionally altered HPDs will not provide adequate attenuation.

## DEMONSTRATING MANAGEMENT SUPPORT

Managers must demonstrate support for HPD utilization and emphasize its importance in several ways:

1. wearing HPDs each time they enter an area with either a voluntary or mandatory HPD wearing policy, no matter for how short a time the managers are in the area,

2. establishing a policy of mandatory HPD utilization,

3. making HPD utilization enforcement part of the basis for performance ratings of the frontline supervisors and production department heads,

4. providing a mechanism for regularly praising or rewarding employees who wear HPDs correctly and consistently,

5. rating the performance of all personnel responsible for any aspect of HPD utilization enforcement, fitting, issuing, or replacement,

6. directing that HPDs be made available to employees for off-the-job noise exposures so that hearing will be protected around the clock,

7. purchasing equipment to allow HPD issuers to monitor the field performance of HPDs as actually used by individuals, and

8. scheduling regular meetings among the HPD fitters and issuers, the reissuers, and the supervisors who enforce utilization on a daily basis.

## REFERENCES

1. Berger, E. H., J. D. Royster, L. H. Royster, and D. Brus. "An Earful of Sound Advice About Hearing Protection." (Indianapolis, IN: E-A-R Division of Cabot Corporation, 1988).
2. Royster, L. H., and J. D. Royster. "Hearing Protection Devices," in *Hearing Conservation in Industry*, A. S. Feldman and C. T. Grimes, Eds. (Baltimore, MD: Williams & Wilkins, 1985).
3. Berger, E. H. "Hearing Protection Devices," in *Noise & Hearing Conservation Manual*, 4th ed., E. H. Berger, W. D. Ward, J. C. Morrill, and L. H. Royster, Eds. (Akron, OH: American Industrial Hygiene Association, 1986).

# Chapter 7

# Audiometric Evaluations

**CHECKLIST FOR AUDIOMETRIC EVALUATIONS**

—Audiometers are in good operating condition.

—Audiometer output level is not adjusted unless it is out of tolerances, and when adjusted both preadjustment and postadjustment readings are permanently recorded.

—Biological (human subjects) calibration and equipment checks are conducted at least weekly during regular testing periods.

—Audiometric technicians use consistent testing methods under professional supervision.

—Technicians instruct employees to listen carefully and respond to the faintest tones they can detect.

—Employees' auditory history information is updated annually and provided to the audiogram reviewer.

—Employees receive immediate feedback from the audiometric technician about audiogram results as related to HPD use.

**continued**

**CHECKLIST continued**

—Employees receive written feedback from the audiogram reviewer about:
1. hearing status compared to normal for age,
2. hearing change over time, and
3. recommendations for better protection on and off the job, or for medical examination or treatment if appropriate.

—The audiogram reviewer looks for significant shifts at any frequency, not just OSHA STS.

—Audiogram reviewers revise employees' reference baseline thresholds for threshold improvement as well as for persistent worsening.

—HCP personnel follow through with counseling and HPD retraining for employees with hearing change.

The audiometric evaluations phase of the HCP ties together all the other phases by indicating whether the program's goal is being achieved: prevention of on-the-job hearing loss. If the HCP is not effective, the result will be seen as worsening hearing thresholds for employees, as well as a higher percentage of the noise-exposed population showing an OSHA STS (standard threshold shift) or the company's significant threshold shift, and an increase in the company's potential liability for compensable occupational hearing loss.

When audiograms detect temporary threshold shift, early permanent threshold shift, or progressive noise-induced hearing loss, the HCP personnel are alerted to take swift actions to halt the loss before the employee's hearing shows a significant deterioration. Because noise-induced hearing loss typically occurs so gradually, the affected individual may not notice the slow change until a large threshold shift has accumulated. Audiometric monitoring can identify individuals who are inadequately protected so that they can be retrained and/or given better HPDs and extra motivational attention to prevent further loss.

However, audiometric evaluations cannot provide reliable data to guide intervention unless they are conducted under adequate quality standards and the results are appropriately evaluated and meaningfully communicated to the employee. For more details about the audiometric phase, see the references.[1-3]

## MANAGEMENT SUPPORT NEEDED FOR QUALITY

Managers must support the audiometric evaluations phase by funding quality services. Because audiometry requires a substantial investment of money and personnel time, it is cost-effective to allocate enough resources to ensure that the desired benefits are obtained from audiometric monitoring. Otherwise, the money is simply wasted on deficient services that do not serve their intended purpose of alerting HCP personnel to take actions that will prevent hearing loss.

### In-House Versus Contracted Services

Managers may choose to contract for employee audiograms to be performed by an external source (a mobile testing contractor or a local clinic). Alternately, management may purchase audiometric equipment and obtain training for internal personnel to perform audiograms under the supervision of an audiologist or qualified physician. The choice depends on factors including the company's philosophy about safety and health, as well as its size and geographical location.

Our experience indicates that the audiometric phase will be *much more effective* in motivating employees if their audiograms are performed and discussed with them by in-plant HCP personnel. If externally contracted services are used, it is *critical* that management assign responsibility to the onsite key individual for making sure that quality services are obtained and for using the audiometric results to motivate employees.

### Well-Qualified Personnel to Perform Audiograms

The audiometric technicians should hold current certification as Occupational Hearing Conservationists from CAOHC (the

Council for Accreditation in Occupational Hearing Conservation, see Resources listing #2). All technicians should use consistent testing methods under the supervision of an audiologist or qualified physician.

### Adequate Time Allowed to Complete Evaluations

The audiogram session cannot achieve its potential for motivating the employee about hearing conservation unless sufficient time is allowed. The technician needs time to obtain auditory history information, inspect HPD condition and fit, properly instruct the employee, carefully administer the audiogram, briefly explain the results to the employee, and document the findings. When the technician is too hurried to do more than give a rapid screening audiogram and herd the employee out the door, the worker correctly perceives that the testing is performed only for OSHA compliance without any sincere interest in protecting his hearing. In this case the employee will usually lose his own motivation to participate in the HCP.

### Obtain All Necessary Information

The OSHA regulations do not require measurement of hearing thresholds at 8000 Hz or the documentation of auditory history information. Auditory history concerns details about the employee's off-the-job noise exposures, medical conditions affecting hearing, the presence of tinnitus (ringing or other noises in the ears), other related information, and a brief visual check of the ear canals. The wise employer will go beyond OSHA requirements to include the 8000 Hz thresholds and history information because these items allow the professional who reviews the audiograms to interpret the results more accurately. The 8000 Hz thresholds help the reviewer in some instances to distinguish between noise-induced hearing loss and age-related hearing change. The auditory history information assists the reviewer in deciding whether hearing changes may be related to on-the-job noise exposures, or whether off-the-job exposures are probably contributing to the hearing loss progression. Access to this information may reduce the number of costly medical referrals made by the reviewer. Additionally, the information can be invaluable in determining the work-relatedness of hearing losses in Workers' Compensation claims.

## Regular Schedule of Audiometric Monitoring

For maximum protection of the company and employees, audiograms should be performed at preemployment, or prior to initial assignment to a noisy work area. Thereafter, audiograms are required annually by OSHA as long as the employee remains in the hearing conservation program (Figure 14). To identify susceptible individuals and detect early hearing deterioration before it progresses significantly, we recommend that audiograms be given every six months during the *first two years* of exposure to TWAs over 99 dB(A). It is also to the employer's advantage to give audiograms when the employee is reassigned out of a noisy job and at the termination of employment to protect against unjustified Workers' Compensation claims.

Many companies have found it is desirable to give audiograms every one to three years as a health screening benefit to employees without on-the-job noise exposure. Annual audiometric results for nonexposed employees also serve as a control group when using audiometric data base analysis (see Chapter 8) to evaluate the HCP's effectiveness.

## Careful Choice of a Professional Reviewer

Management should ensure that the program supervisor who reviews employees' audiograms is a well-qualified professional with specific training and experience in the area of industrial hearing conservation. Such a choice will benefit the employer as well as the employees, as an experienced audiologist or physician is less likely to mistake nonoccupational loss as being job-related.

## Feedback and Follow-Up

Because the audiometric session is the greatest opportunity to motivate employees concerning hearing conservation, management should allow an extra two minutes for the audiometric technician to give the employee simple, brief remarks about his hearing status immediately after the audiogram is completed to praise the worker or warn him that better HPD utilization is needed.

**Figure 14.** Annual audiometric evaluations monitor the employee's hearing thresholds. If noise-related hearing changes are detected, counseling and HPD retraining may help the employee achieve better protection.

*All* employees (not just those with threshold shifts), should receive written feedback from the professional reviewer. When the reviewer points out potentially noise-induced hearing changes, company personnel need to take decisive follow-up actions. These actions include individual counseling, refitting and retraining for HPD utilization, encouraging the employee to wear HPDs during off-the-job noise exposures, and more careful supervision of on-the-job HPD use.

When the reviewer indicates that the shifts appear unrelated to noise exposure, the employee should be urged to seek otological/audiological evaluation and treatment. Occasionally the reviewer may recommend that the company pay for an otological evaluation if it appears that a medical ear condition is caused or aggravated by the use of HPDs on the job.

If audiograms are filed and forgotten rather than being used to guide follow-up actions, then the audiometric phase simply documents hearing loss rather than helping to prevent it. Feedback to the employees about their audiometric results is the best tool available to motivate individuals to protect their own hearing.


## QUALITY CONTROL RESPONSIBILITIES
## OF AUDIOMETRIC TECHNICIANS

The accuracy and usefulness of the audiogram results depend on the care and attentiveness of the audiometric technicians.

### Maintaining Test Equipment and Environment

It is critical for audiogram accuracy that the audiometric technicians perform and document daily calibration checks and self-listening checks of audiometer function. If an electro-acoustic blockhead device is used to check calibration of audiometer output levels, the technician must still *listen through the earphones* for signal distortion or erratic responses which the device cannot detect. A logbook of biological and/or electroacoustic output checks and listening checks of audiometer function is strongly recommended.

In order to measure thresholds accurately for employees, at least down to 0-dB hearing levels (note that some employee's thresholds would be as low as − 15 dB if your equipment and environment is able to test this range), and to ensure that company audiograms would be acceptable for legal purposes (such as in Worker's Compensation claims), the test room must be quiet enough to meet ANSI standard S3.1–1977. Note that this standard requires lower background levels than the OSHA Hearing Conservation Amendment. The technician should check and document background levels periodically.

Acoustic and exhaustive audiometer calibrations should be scheduled regularly, but calibration services should not be allowed to adjust the audiometer unless it fails to meet calibration tolerances. Unnecessary annual adjustments typically add see-saw variability to the audiometric data, interfering with the interpretation of both individual and group hearing trends. If audiometer adjustments must be made because tolerances are exceeded, the calibration company must be required to provide both preadjustment and postadjustment measurement values so that the size and direction of changes are documented, and appropriate correction to the audiometric data can be made before making important decisions as to employee referral or program effectiveness.

To prevent another source of measurement variability, the same audiometers should be used consistently rather than switching back and forth between types of audiometers (manual, self-recording, and microprocessor audiometers).

### Using Consistent Instructions and Testing Methods

All audiometric technicians should use the same standardized testing method under the supervision of the audiologist or qualified physician. When technicians differ in their response instructions to employees or their signal presentation and threshold recording methods, the excess variability in threshold measurements interferes with interpretation of hearing trends.

It is very important to instruct employees to listen carefully and respond to the faintest tones they can detect—rather than waiting until the tones are loud and clear before responding. (The wording of the instructions can be adjusted to whatever vocabulary seems most meaningful to the employees, but this same concept of listening for faint tones must be communicated.)

Because hearing thresholds may improve over time, the audiometric technician must try to measure the best thresholds possible for the employee, rather than simply duplicating the results of the last test.

## Maintaining Complete Records

The audiometric record should indicate the specific equipment used, calibration date, the name of the technician, the date and time of the test, the threshold values obtained, the technician's judgment of the subject's response reliability, the HPD inspection results and refitting or reissuing record, the employee's TWA, and the technician's comments, if any. The professional reviewer's recommendations should also be recorded, as well as documentation that follow-up actions were carried out.

## SCHEDULING AUDIOGRAMS

The audiometric evaluations should be scheduled in close coordination with production supervisors in order not to disrupt production more than necessary. If supervisors' needs are ignored, the supervisors will be less likely to fulfill their HCP responsibility of monitoring HPD utilization. To minimize the disruptive effect on departments and to ensure that employees who are transferred among departments do not accidentally miss their annual audiograms, it is often useful to schedule audiograms in the employee's birth month.

Participation in annual audiometric evaluations should be a condition of employment, not a choice.

The baseline audiogram should be administered when the employee has not been exposed to noise for the preceding 14 hours. However, annual audiograms should be administered *during* the workshift, not before it, so that the results will detect any temporary threshold shifts in hearing that may be occurring in employees who are not adequately protected. Some companies go to great pains to perform audiograms before employees begin their workshifts so that no temporary threshold shifts will be measured. However, we *need* to be able to detect the presence of temporary threshold shifts in the population's audiometric data in order to identify those employees who may need better protection and instruction and to identify an ineffective HCP before the population develops permanent hearing loss.

## USING THE AUDIOGRAM SESSION TO BEST ADVANTAGE

The annual audiogram is often the only predictable time during the year when the individual employees have the opportunity for a one-to-one conversation about hearing conservation and their own hearing status. This time is the best chance to motivate workers to protect their hearing. Whether the technicians are external contract personnel or internal staff members, they absolutely must demonstrate enthusiasm for hearing conservation and sincere interest in the employee while performing audiograms and giving feedback to employees (Figure 15).

### Auditory History Information

OSHA regulations do not require maintenance of auditory history information (documentation of past noise exposures in the military and previous jobs, off-the-job noise exposures such as gunfire or power tool use, and ear-related medical history). However, it is in the employer's best interest to record these data to allow for better interpretation of the employee's audiometric results and for protection against future hearing loss compensation claims. Updated history information allows the professional reviewer to make recommendations with more insight and confidence concerning probable causes for threshold shifts. In addition, the history questions remind the employee that off-the-job factors influence his hearing status.

### Checking HPDs at Time of Audiogram

As previously discussed in Chapter 6, the audiometric technician should check the condition and fit of the employee's HPDs, change them if needed, and reinstruct the wearer.

**Figure 15.** Immediate feedback about the audiogram results can motivate the employee to protect his hearing on and off the job.

## Immediate Feedback About Hearing Trends

The professional who supervises the audiometric technician can provide guidelines for saying a few sentences about employees' hearing as soon as they step out of the audiometric booth, when they are most interested in their hearing and most receptive to suggestions (Figure 16).

If hearing thresholds have remained stable, praise the employee for a job well done! If thresholds appear to be worsening, advise the employee of the need to determine the causes of

# Normal Audiogram
# and Degrees of Hearing Loss

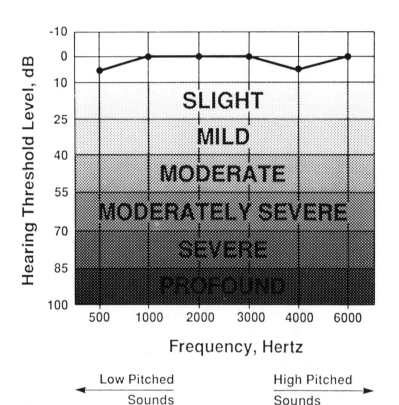

**Figure 16.** The audiogram shows hearing thresholds for tones at different pitches or frequencies. Normal hearing thresholds fall within the *unshaded area* of the chart. If noise-induced hearing damage or age-effect hearing loss begins to develop, the thresholds for the higher frequencies (3000 to 6000 Hertz) will start to fall into the shaded areas indicating degrees of hearing loss.

the hearing change, such as medical problems or inadequate use of HPDs on or off the job. Note that the professional reviewer will provide more detailed written feedback about the hearing changes.

## FOLLOWING UP ON AUDIOMETRIC RESULTS

Every audiometric record must be reviewed to classify the hearing trends and determine whether other actions are needed.

### Careful Audiogram Review

The supervising professional may set up procedures for the technician to follow to review routine records (those with normal hearing and no shifts or improvements) and prepare employee feedback notices. Alternately, it may be easier for the company to have the professional review the routine records as well as the problem audiograms. The reviewer needs to look for threshold improvements and significant threshold shifts at any test frequency (not just OSHA standard threshold shifts). Significant shifts would include a change of 15 dB at any frequency that is confirmed as being persistent on a retest (or the next regular annual test), or an unconfirmed change of 20 dB at any frequency. The professional should also check for audiogram patterns indicative of possible otological pathology.

---

ALL EMPLOYEES NEED FEEDBACK,

NOT JUST THOSE WITH OSHA STS!

---

### Prompt Meaningful Feedback

The reviewer (or technicians under professional supervision) should provide each employee written feedback describing the worker's hearing status in terms of three aspects:

1. comparison to the expected age-effect hearing levels for the worker's age/sex/race group and the hearing ability needed for unimpaired communication,

2. a description of the amount of change seen in the current audiogram compared to past results and the designated reference baseline, and

3. recommendations, including praise for stable hearing, warnings to use HPDs more carefully on and off the job if hearing changes are observed, suggestions to seek medical attention or further audiological evaluation, etc.

---

AUDIOGRAMS DON'T PREVENT HEARING LOSS,

BUT FEEDBACK AND FOLLOW-UP CAN!

---

### Taking Appropriate Follow-Up Actions

OSHA regulations specify required follow-up actions for OSHA STS (standard threshold shift, which is a change of 10 dB in the average thresholds at 2000, 3000, and 4000 Hz in either ear after optional corrections for aging). However, in effective HCPs, aggressive follow-up for beginning hearing shifts will prevent losses from progressing into OSHA STSs.

In an excellent HCP, the employee with beginning shifts will receive not only a written confirmation from the professional reviewer, but also face-to-face counseling from onsite HCP personnel based on the reviewer's comments, reevaluation of HPD adequacy, retraining in HPD placement, and extra supervision in on-the-job HPD utilization. A retest audiogram may be given to see if the shifts persist or disappear. Individuals with potential medical ear conditions will be counseled by the employer to seek medical evaluation and treatment, or possibly sent for treatment at company expense.

### EDUCATING EMPLOYEES TO TAKE RESPONSIBILITY

As employees become familiar with the audiometric evaluation process, understand their audiogram results, and learn how their everyday habits in HPD utilization can affect their hearing trends, they will be encouraged to take the lead in protecting their own hearing on and off the job. No matter how much effort the employer puts into the HCP, it is impossible to inspect whether each employee is wearing HPDs adequately each day. Eventually the individual worker must accept responsibility for following through with good hearing protection habits. The best tool the hearing conservationist has to motivate individuals about protecting their hearing is to counsel them about their audiogram results.

## REFERENCES

1. Royster, J. D. "Audiometric Evaluations for Industrial Hearing Conservation," *Sound and Vibration* 19(5) 24–29 (1985).
2. Morrill, J. C. "Hearing Measurement," in *Noise & Hearing Conservation Manual*, 4th ed., E. H. Berger, W. D. Ward, J. C. Morrill, and L. H. Royster, Eds. (Akron, OH: American Industrial Hygiene Association, 1986).
3. Miller, M. H. *Occupational Hearing Conservation Training Manual*, 2nd ed. (Springfield, NJ: Council for Accreditation in Occupational Hearing Conservation, 1987).

# Chapter 8

# Making Sure That the HCP Works

---

**CHECKLIST FOR HCP EVALUATION**

—There is a key individual overseeing all five phases of the HCP.

—HCP team members check that all tasks are accomplished and documented.

—HPDs are potentially effective in actual use.

—HPD utilization is enforced.

—Active communication is maintained among HCP team members and with all personnel up and down the company hierarchy.

—Management holds personnel accountable for their HCP performance and gives praise or criticism as appropriate.

—Audiometric data base analysis is used to evaluate the HCP's effectiveness in preventing on-the-job hearing loss.

---

## THE TEAM APPROACH

Now that you have reviewed all five phases of HCPs, take another look at Figure 2 and consider the importance of teamwork in achieving the goal of preventing hearing loss on the job. No single phase of the HCP can work effectively in isolation from the others. Many people are involved: foremen, tool crib clerks, safety officers, audiometric technicians, nurses, personnel directors, industrial hygienists, engineers, audiologists, physicians, and others. No single discipline can claim superior ability to run the HCP: the program depends on the cooperation of many people, under the leadership of the key individual. The key individual can be anyone with the interpersonal skills and managerial know-how to coordinate the contributions of these diverse personnel toward preventing hearing loss. The key individual is the catalyst who makes the HCP work by maintaining communication among the team members to achieve a unified HCP.

### Documentation and Recordkeeping

When the OSHA inspector visits, the only way the company can demonstrate that the HCP is satisfactory, for OSHA purposes, is through adequate documentation of the five phases. Some people actually refer to recordkeeping as a sixth phase, but recordkeeping is an integral part of each phase, not a separate activity. For example, the audiometric phase depends on cumulative records to show hearing changes over time, plus records of auditory history information, employee noise exposure, and HPD fitting and reissuing to evaluate whether threshold shifts may be work-related.

In addition to the legal documentation it provides, good recordkeeping is helpful in monitoring the program and sharing information among HCP team members. The key individual should assign responsibility to HCP team members for maintaining the records associated with their duties and should ensure that the records are accessible. A list of the records that are needed is shown as Table 4, and further details are provided in more comprehensive references.[1-2]

**Table 4.   Documentation Guidelines for the HCP**

A. RECORDS SPECIFIED IN OSHA'S HEARING CONSERVATION
   AMENDMENT

1. Noise exposure measurements
   Detailed survey report must include complete list of instruments used
      in survey, calibration, measurement positions, tables of sound level
      measurements, TWA calculations, etc.
   List departments or employees with TWAs of 85 dB(A) or over.
   For OSHA, retain 2 years or until a new sound survey.
   Keep indefinitely for Worker Compensation purposes.

2. Documentation of engineering/administrative noise controls, including:
   results of engineering sound surveys
   installations completed and noise reduction achieved
   regular maintenance of machinery and controls

3. Documentation of annual educational programs, including:
   content of presentation
   names of presenters
   list of employees who attended

4. Documentation of hearing protection phase of HCP:
   date of initial HPD fitting of each employee
   brand and size of HPD fitted (in each ear if appropriate)
   employee's signature for training in HPD use and care
   documentation of employer's supervision of correct HPD use such
      as walk-through checks of utilization, etc.
   NRR and TWA calculations showing HPD adequacy

5. Employee's audiometric records, including:
   name, age, job classification, and TWA exposure
   date of audiogram and name of audiometric technician
   audiometer model/SN and date of its last calibration
      Retain for duration of employment for OSHA.
      Keep indefinitely for Worker Compensation.

6. Supporting records for audiometric phase of HCP:
   technician's certification credentials
   audiometer make, model, and serial number
   audiometer acoustic and exhaustive calibration records
   biological calibration check records of audiometer
   background sound levels in audiometric test room

7. Documentation of audiogram review and follow-up actions:
   review of each audiogram by professional or technician
   credentials of audiologist or physician reviewer
   reviewer's follow-up recommendations
   documentation that employer recommended follow-up

documentation of employee's written notification of STS
employee's signature indicating OSHA STS follow-up
documentation of HPD utilization enforcement after STS

B. ADDITIONAL RECORDS EMPLOYER SHOULD KEEP FOR THE HCP

1. Audiometric Phase
   auditory history information for each employee in HCP
   annual history updates
   annual otoscopic checks
   preemployment or preexposure audiograms
   termination audiograms

2. Hearing Protection Phase
   dates of HPD reissuing, brand and size reissued
   annual documentation at audiogram time that:
      employee's HPD is correct size, in good condition
      employee can demonstrate proper use of the HPD
   list of HPDs the employer allows to be used in work environments
   with different TWA ranges, considering real-world attenuation
   (must de-rate the NRR)

3. Sound Exposure Monitoring
   noise map of the facility, showing:
      HPD-required areas
      [TWAs = 90 dB(A) or higher]
      areas where only certain HPDs are acceptable
      [TWAs = 95 dB(A) or higher]
      areas where HPDs are optional
      [TWAs below 90 dB(A)]

RECORDING HEARING LOSS ON THE OSHA 200 FORM

OSHA has not yet set forth any guidelines on the recording of occupational
hearing loss on the Form 200 as an occupational illness (for gradual loss) or
injury (for sudden traumatic loss). However, several professional associations
(AIHA, ASHA, NHCA), have supported the recording of a confirmed OSHA
STS that has been judged as occupationally caused or aggravated (see D. P.
Driscoll and J. C. Morrill, *American Industrial Hygiene Association Journal* 48:
A714-A716 (1987).

## EVALUATING YOUR HCP

The most basic form of program assessment is to use the check-
off lists provided in this booklet to see whether all the bases have
been covered. Have all the tasks been accomplished? For example, the key individual might discover that a group of employees

did not attend an educational program or missed their annual audiograms. Just as important, have the more subjective guidelines been met? For instance, has the key individual made a point to get feedback from the HPD reissuers and employees concerning the practicality of the new earplug?

The checklist approach is fine as far as it goes, but it does *not* assess the effectiveness of the program, merely its completeness. To evaluate whether hearing loss is really being prevented, a different approach is needed.


## AUDIOMETRIC DATA BASE ANALYSIS (ADBA)

The audiometric results for the noise-exposed employees provide the only objective indication of whether the HCP is succeeding in preventing occupational hearing loss. Reviewing the audiometric records for employees one at a time detects hearing changes for individuals, but it does not provide an overall picture of how well the group of workers is being protected. In contrast, analysis of group audiometric data can show the trends for departments and the whole plant.

Over the past several years, the ANSI S12–12 Working Group for Evaluation of Hearing Conservation Programs (see Resources listing #5) has been developing simple procedures which the employer can apply to the audiometric data for the group of exposed employees to assess whether they are being adequately protected. The advantage of audiometric data base analysis (ADBA) procedures is that they can detect problems in the HCP early, before individual workers develop significant permanent hearing loss. If the results show that the HCP is ineffective, the employer can improve program practices to prevent additional hearing loss from developing.

AUDIOGRAMS DO NOT PREVENT HEARING LOSS,

BUT USING ADBA RESULTS CAN!

The ANSI S12-12 Working Group has recently issued its recommendations for procedures to use in evaluating the effectiveness of HCPs. The general guidelines and two of the recommended ADBA procedures are summarized below, and detailed discussions are available.[3-5]

### Audiometric Variability as an Indicator

The procedures recommended for audiometric data base analysis are based on the year-to-year variability in audiometric thresholds. If you look at the audiogram results for a person from one year to the next, thresholds at some frequencies may be a little better, while thresholds at other frequencies may be a bit worse. Variability in audiometric threshold measurements comes from three main sources:

1. normal fluctuations in the responsiveness of the person being tested (unavoidable variability),
2. inconsistencies in the equipment and testing methods used to administer the audiogram (avoidable measurement error),
3. true threshold changes due to temporary or permanent hearing loss if employees receive inadequate protection from noise (what the HCP is trying to prevent).

ADBA looks at the total variability in employees' hearing threshold measurements. If the variability in the HCP is in the same low range that is achieved in low-noise-exposed or non-exposed industrial groups, then the HCP is judged to be successful in avoiding measurement error (2) and in preventing hearing loss (3). However, if variability is too high, then the hearing conservationist must determine whether the cause is a problem with the audiometric testing procedures or a real indication of developing hearing loss resulting from inadequate protection. Until the variability is reduced, the HCP is judged ineffective, because unreliable audiometric results may not be capable of identifying actual hearing loss in individuals.

### Recommended Guidelines

Two variability procedures recommended by the ANSI S12-12 Working Group are based on counting the percentage of employees whose hearing shows changes of 15 dB or more between two sequential (consecutive) annual audiograms, such as from test 1 to test 2, from test 2 to test 3, etc. (Note! the 15 dB shift referenced here simply serves as a flag and is not directly related to the 15 dB significant threshold shift criterion referenced earlier). Threshold changes are counted both toward better hearing and toward worse hearing to yield values for these two ADBA procedures:

1. Percent Worse Sequential (%Ws): the percentage of employees who show a worsening of 15 dB or more in thresholds for at least one test frequency (500 Hz through 6000 Hz) in either ear between two sequential audiograms,

2. Percent Better or Worse Sequential (%BWs): the percentage of employees who show *either* an improvement *or* a worsening of 15 dB or more in thresholds for at least one test frequency (500 Hz through 6000 Hz) in either ear between two sequential audiograms.

Based on applying these procedures to the audiometric data for over 20 industrial HCPs, the ANSI S12-12 Working Group has defined ranges of that which indicate the HCP's quality as acceptable, marginal, or unacceptable. These ranges are shown in Table 5. The ranges are slightly different for the first four years of audiometric testing (sequential test comparisons 1-2, 2-3, and 3-4) than for later years of testing. Note that before the procedures

Table 5. HCP Effectiveness Classification and Corresponding Recommended Value Ranges for Two ADBA Procedures Applied to Sequential Test Comparisons with No Age Corrections: Percent Worse (%Ws) and Percent Better or Worse (%BWs)

| | Value Range, Percent | | |
| | Over First Four Test Comparisons | Over Later Test Comparisons | |
| HCP Rating | %Ws | %Ws | %BWs |
| --- | --- | --- | --- |
| Acceptable | < 20 | < 17 | < 26 |
| Marginal | 20 to 30 | 17 to 27 | 26 to 40 |
| Unacceptable | > 30 | > 27 | > 40 |

are applied, the population must be restricted to a group of workers who all have a specified number of audiograms (see references 4–5 for more detail).

### Advantages and Benefits of ADBA

By using the variability procedures to evaluate the audiometric data, HCP personnel can detect problems in the HCP quickly—within the time period of two tests—and then act to correct the deficiencies *before* many employees develop significant permanent hearing changes. In other words, the analysis helps HCP personnel to *prevent* hearing loss (Figure 17).

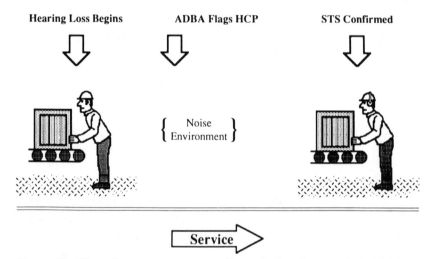

Hearing Loss Begins        ADBA Flags HCP              STS Confirmed

{ Noise Environment }

Service

**Figure 17.** When the company uses audiometric data base analysis (ADBA), problems in the HCP can be detected *before* employees develop significant permanent hearing threshold shifts.

The findings from applying ADBA are converted into simple bar graphs that provide useful feedback for the supervisors and employees in different departments to show how better HPD utilization can result in a more effective HCP, or that an effective HCP is in place.[3] Such concrete results are an effective way to motivate employees.

The key individual can use ADBA procedures to guide policy decisions. If HCP personnel are unsure whether a particular HPD provides adequate protection, the ADBA results for employees wearing different HPDs can be compared. Similar department comparisons can show whether required HPD utilization is needed to protect employees in a low-noise department.[5]

Finally, ADBA findings give objective evidence to demonstrate for management when the budget allocations for the HCP need to be increased or redistributed in order to improve protection in departments with poorer performance. After HCP changes have been implemented, ADBA results show management the associated gains: less hearing loss and reduced potential liability for Worker's Compensation claims.

## REFERENCES

1. Stewart, A. P. "The Comprehensive Hearing Conservation," in *Hearing and Conservation in Industry, Schools, and the Military*, D. M. Lipscomb, Ed. (Boston, MA: College-Hill Publications, 1988).
2. Gasaway, D. C. *Hearing Conservation: A Practical Manual and Guide.* (Englewood Cliffs, NJ: Prentice-Hall, 1985).
3. Royster, L. H., and J. D. Royster. "Making the Most Out of the Audiometric Data Base," *Sound and Vibration* 18(5) 18–24 (1984).
4. Royster, L. H., and J. D. Royster. "Getting Started in Audiometric Data Base Analysis," *Seminars in Hearing* 9:325–338 (1988).
5. Royster, J. D., and L. H. Royster. "Audiometric Data Base Analysis," in *Noise & Hearing Conservation Manual*, 4th ed., E. H. Berger, W. D. Ward, J. C. Morrill, and L. H. Royster, Eds. (Akron, OH: American Industrial Hygiene Association, 1986).

Chapter 9

# Reducing Your Compensation Liability for Noise-Induced Hearing Loss

---

**CHECKLIST FOR PREVENTING COMPENSATION CLAIMS**

—ADBA results show the HCP is effective.

—Preemployment audiograms are required.

—Postemployment audiograms are mandatory.

—Annual audiograms are given to employees with TWAs of 85 to 99 dB(A).

—For employees with TWAs of 100 dB(A) or higher, audiograms are given every six months for the first two years of exposure.

—Significant threshold shifts are identified and receive follow-up action to halt hearing loss.

—Documentation is kept of HPD fitting, issuing, replacement, and user training for employees.

—Auditory history information for employees is obtained and updated annually.

**continued**

---

---

**CHECKLIST continued**

—When a hearing loss claim is filed, one check on its validity is obtained using the ISO 1999.2 model.

—All company personnel, consultants, attorneys, and medical experts are adequately educated in preparation for the hearing or trial.

---

### EFFECTIVE HCPs—NO CLAIMS FOR PREVENTABLE HEARING LOSS

In industrial work environments, hearing loss occasionally occurs due to unforeseeable accidents that create immediate permanent hearing loss due to mechanical damage to the inner ear by an intense pressure wave (acoustic trauma). This may happen if the employee is exposed to a blast or explosion in a work setting where HPDs are not required, or during a time period when an employee has temporarily broken the seal of HPDs for adjustment. Traumatic hearing loss may also occur in an accident involving a blow to the head. These types of immediate hearing losses are classified as occupational injuries for purposes of Workers' Compensation. Fortunately, they are rare.

By far the most common occupational hearing losses develop slowly over time, when the worker is exposed to harmful levels of noise each day without being adequately protected. Under Workers' Compensation regulations, noise-induced permanent threshold shift (NIPTS) which develops gradually is classified as an occupational illness.

It is not reasonable to expect management to be able to completely eliminate compensable hearing losses resulting from accidental acoustic trauma. However, in the case of hearing loss resulting from daily noise exposures over an extended time period, today there is no justification for a company to allow an employee to develop a compensable hearing loss on the job without being aware of it.

For at least the past 20 to 30 years we have had sound survey instrumentation to identify noise-hazardous work areas, HPDs that (singly or in combination) can provide adequate protection in industrial noise environments exhibiting daily OSHA TWAs up to at least 100 dB(A), audiometers to monitor the employee's hearing over time, and adequate audiogram review procedures to identify noise-susceptible employees before their hearing changes significantly. Finally, today we have the audiometric data base analysis (ADBA) criteria to identify ineffective HCPs before significant hearing damage occurs in the noise-exposed population. This warning from ADBA results allows prudent management to implement corrective actions in time to prevent compensable hearing loss.

> IF AN EMPLOYEE DEVELOPS
>
> A COMPENSABLE HEARING LOSS
>
> DUE TO ON-THE-JOB NOISE,
>
> THE EMPLOYER SHOULD PAY!

## HOW TO CALCULATE COMPENSATION LIABILITY

The rules for determining compensation for occupational hearing loss vary widely among different jurisdictions.[1] Each state is a separate jurisdiction which establishes its own regulations for Workers' Compensation, while employees in federal, military, and other selected occupations (such as railroad employees and longshoremen) are covered under their own regulations. In some jurisdictions occupational hearing loss is not compensated, or only certain types of occupational loss are compensable (such as complete loss of hearing, or traumatic loss through injury as opposed to gradual loss).

Of the jurisdictions which do allow compensation for occupational hearing loss, most use some type of formula for determining whether a hearing loss is potentially compensable and for setting the dollar value of compensation according to the percent of assumed disability. Typically the hearing loss formulas involve:

1. consideration of the average of hearing thresholds at only selected audiometric frequencies (such as 0.5, 1, and 2 kHz, or 0.5, 1, 2, and 3 kHz, or 1, 2, and 3 kHz),

2. differential weighting of the better ear and poorer ear thresholds in determining a bilateral average threshold for the selected frequencies (usually the better ear is given five times the weight of the poorer ear),

3. a *low fence* for the bilateral average threshold at which potentially compensable hearing disability is assumed to begin (typically 25 dB), and a *high fence* for the bilateral average threshold (typically 92 dB) at which the disability is assumed to be complete (worth 100% of the maximum award), and

4. a dollar value of the maximum award possible for 100% assumed disability.

The combination of items 1, 2, and 3 is typically referred to as a *formula* for compensable hearing loss. To illustrate how these formulas work in one example, consider the hearing threshold levels that are shown in Table 6.

**Table 6.   One Employee's Hearing Threshold Levels, dB**

| Left Ear | | | | | | Right Ear | | | | | |
|---|---|---|---|---|---|---|---|---|---|---|---|
| Audiometric Test Frequency, kHz | | | | | | | | | | | |
| 0.5 | 1 | 2 | 3 | 4 | 6 | 0.5 | 1 | 2 | 3 | 4 | 6 |
| 19 | 24 | 46 | 66 | 73 | 74 | 17 | 23 | 43 | 64 | 70 | 70 |

The percent of assumed disability for the hearing loss and the corresponding dollar value are calculated through the following steps:

1. Average the employee's hearing threshold levels separately in each ear for the desired audiometric frequencies. For this example we will use 0.5, 1, and 2 kHz. The average thresholds for these frequencies for the audiogram shown in Table 6 are 29.7 dB for the left ear [HLl] and 27.7 dB for the right ear [HLr].

2. Next, weight the values for the two ears (HLl and HLr) as follows: add 5 times the HL value for the better-hearing ear, plus 1 times the HL value of the poorer-hearing ear, and divide the sum by 6. For this employee the combined hearing level (HLc) is [(5 × 27.7) + 29.7]/6 = 28 dB.

3. Determine the number of decibels by which the HLc exceeds the low fence. Using the typical low fence value of 25 dB, the HLc of 28 dB exceeds the low fence by 3 dB (28 − 25 = 3).

4. Determine the percentage of assumed disability. If the low fence is 25 dB and the high fence is 92 dB, there is a difference of 67 dB between 0% disability and 100% disability, so each decibel by which HLc exceeds the low fence is approximately 1.5% disability [100%/67dB = 1.5%/dB]. For our example, the HLc of 28 dB yields 4.5% disability [3 dB × 1.5%/dB = 4.5%].

5. Determine the dollar value. Assuming for this example that the jurisdiction pays a maximum of $100,000 for 100% hearing disability, this employee would receive 4.5% × $100,000, or $4,500.

The hearing loss formulas most commonly used by state jurisdictions involve hearing thresholds averaged over 0.5, 1, and 2 kHz (as illustrated above) or 0.5, 1, 2, and 3 kHz, both with a low fence of 25 dB and a high fence of 92 dB. The federal government utilizes a formula that averages the thresholds at 1, 2, and 3 kHz, with the same fences. The rationale for the selection of audiometric frequencies that should be included in a formula for Workers' Compensation is discussed elsewhere.[2] The choice of audiometric frequencies included in the formula critically affects the assumed percentage of disability for any individual employee. To illustrate this point, look at Table 7 and note the differences in the disability percentage and dollar value if three different hearing loss formulas are applied to this same employee's thresholds.

Table 7.   HLc Values, Disability (For a Low Fence Set at 25 dB), and Payment in Dollars Based on a Maximum Award of $100,000

| Frequency Combination, kHz | HLc, dB | Disability, Percent | Payment, Dollars |
|---|---|---|---|
| 0.5,1,2 | 28.0 | 4.5 | 4,500 |
| 0.5,1,2,3 | 37.1 | 18.1 | 18,100 |
| 1,2,3 | 43.7 | 28.0 | 28,000 |

The hearing thresholds shown in Table 6 are representative of the expected hearing loss for highly noise-susceptible individuals after 35 years of exposure to a daily TWA of 95 dB(A). For the degree of hearing loss shown and the resulting problems in daily living which an individual with this hearing loss would experience, the potential compensation awards as indicated in Table 7 are small, even when the frequency combination of 1, 2, and 3 kHz is utilized.

More important, this hearing loss took 35 years to progress to its present degree of severity, even for these very noise-susceptible individuals. In an effective HCP, the developing hearing loss of these employees would have been detected and corrective actions taken. In addition, the problems that allowed hearing loss to begin would have been solved, thereby minimizing the potential for the hearing loss to progress to a compensable level.

## HOW A COMPENSABLE HEARING LOSS DEVELOPS

### A Model for Prediction

The hearing ability of an employee population can be thought of as the combination of two parts:

1. a component of occupational noise-induced permanent threshold shift (NIPTS), and

2. a component that includes all other factors combined (age-related loss plus medical hearing problems and non-occupational noise exposure) and is termed AGE EFFECT.

International Standard ISO 1999.2 [3] provides a model for estimating the hearing ability of a population, using the following equation:

$$\text{HEARING LEVEL} = \text{NIPTS} + \text{AGE EFFECT} - [(\text{NIPTS} \times \text{AGE EFFECT})/120]$$

The term in brackets that is subtracted away in this equation is a correction factor which prevents the total predicted hearing level from reaching unmeasurably high thresholds if extremely high values exist for either the NIPTS or AGE EFFECT components. If an employee comes to work with a large preexisting NIPTS component, then his or her hearing cannot worsen as rapidly due to AGE EFFECT as an employee who started work with normal hearing, assuming all other factors are equal. Alternately, if it is assumed that an older employee who already has significant age-effect hearing loss begins to be exposed to noise for the first time, his hearing cannot deteriorate from NIPTS as rapidly as would occur in a younger employee with initially good hearing.

**Predicting Compensable Loss**

In the ISO 1999.2 hearing loss equation shown above, the AGE EFFECT and NIPTS values vary for different audiometric test frequencies, as well as for the population's susceptibility to noise. The hearing levels for each audiometric frequency must be predicted separately, and they show different rates of change over years of exposure. In the example that follows, we have combined the ISO 1999.2 predictions for separate audiometric frequencies in order to illustrate the rates of hearing change for the three frequency averages which we previously discussed as the basis for many Workers' Compensation formulas: 0.5, 1, and 2 kHz, 0.5, 1, 2, and 3 kHz, and 1, 2, and 3 kHz.

Let's take a population of employees hired with normal hearing at age 20 years, then exposed to a daily $L_{eq}$ of 95 dB(A) for 50 years of employment. To estimate the maximum hearing change over time, assume that the population represents the 5th percentile in terms of susceptibility to both noise and age effects. (Only 5% of the population would develop the same or more hearing loss than those in the 5th percentile.) The expected changes in the hearing levels at the three frequency averages are shown in Figures 18, 19, and 20 for the assumed AGE EFFECT alone, NIPTS alone, and total hearing loss.

In Figure 18 the AGE EFFECT component of the population's hearing starts off at 10 dB without prior noise exposure because we selected the 5th percentile, which has poorer-than-average hearing throughout life. [Note: the 5th percentile of the screened age-effect population shown here has similar age-effect hearing levels as the 50th percentile of an unscreened population, which is somewhat typical of the general U.S. population without on-the-job noise exposure.][4-6]

Observe in Figure 18 how the rate of hearing decline from the AGE EFFECT *increases* with time as the population ages. In contrast, in Figure 19 the NIPTS component shows the most rapid rate of hearing decline during the earliest years of exposure, with *decreasing* progression of NIPTS over time. For the total hearing loss (HLc), the combination of NIPTS and AGE EFFECT as shown in Figure 20, the rate of hearing decline over time is more constant and slower than the maximum rates of change for either of the separate components. The separate AGE EFFECT and NIPTS components, which act most strongly at opposite ends of the employee's career, combine to produce a relatively uniform rate of hearing decline over time.

### Implications for Workers' Compensation

In terms of Workers' Compensation liability, we can make several important observations from the data presented in Figures 18, 19, and 20. *First*, even though we assumed a population with very high susceptibility to both NIPTS and AGE EFFECTS, and even though we assumed a high TWA noise exposure value, it still takes around 7 years of noise exposure before these employees develop enough hearing loss to become potentially

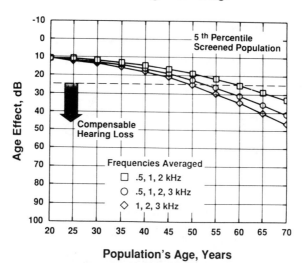

**Figure 18.** Averaged AGE-EFFECT component of hearing level using three compensation hearing loss formulas as a function of age for susceptible ears (5th percentile of a screened presbycusis population).

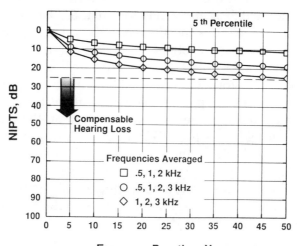

**Figure 19.** Averaged NIPTS component of hearing level using three compensation hearing loss formulas as a function of exposure duration at an $L_{eq}$ of 95 dB(A) for susceptible ears (5th percentile).

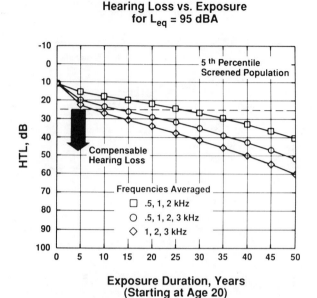

**Figure 20.** Averaged total hearing level (HLc) using three compensation hearing loss formulas as a function of age and exposure duration at an $L_{eq}$ of 95 dB(A) for susceptible ears (5th percentile).

compensable using the 1, 2, and 3 kHz formula, and about 27 years using the 0.5, 1, and 2 kHz formula. *Second*, since the total hearing change is due mostly to NIPTS during the early years and due mostly to AGE EFFECTS during the later years, it is critical to control the NIPTS component early in the employees' noise exposure history. Otherwise, if the NIPTS component has been allowed to develop, then management has lost the opportunity for controlling company liability for on-the-job noise-induced hearing loss, since AGE EFFECT hearing decline is inevitable and will add to the total hearing loss.

The bottom line is that management, in general, has more than adequate time to detect hearing loss due to noise exposure and, if desired, make the necessary corrections in the HCP before a compensable hearing loss develops.

### Age Corrections Are Inappropriate

Since hearing levels do decline with age, why don't we correct the employee's HTLs for normal aging before applying a formula for calculating compensability? Observe in Figure 18 that the AGE EFFECT component, even for this most susceptible 5th percentile of the population, does not exceed the 25 dB fence for the average thresholds at 0.5, 1, and 2 kHz until around age 60.

Remember that 95% of the population would develop age-related hearing loss *more slowly* than the 5th percentile. The vast majority of non-noise-exposed individuals would *never* reach the low fence for any hearing loss compensation formula during their working lifetime. The low fence itself acts as a crude correction for age-related hearing loss, since the typical employee would never exceed the low fence without developing NIPTS. Therefore, it is inappropriate to use any additional age-effect correction.

## PREVALENCE OF POTENTIALLY COMPENSABLE HEARING LOSS IN INDUSTRY

In the preceding sections we utilized three different formulas for averaging hearing thresholds to calculate the weighted binaural average threshold, or HLc. Now let's look at the prevalence of potentially compensable hearing loss in U.S. industries by showing the distribution of HLc values for employees in typical industrial populations, using the same three hearing loss formulas.

Presented in Table 8 is a summary of the findings from applying the three previously defined hearing loss formulas to the audiometric data bases made available to the ANSI S12–12 Working Group for Evaluation of Hearing Conservation Programs. The average percentages of employees compensable using each of the two most common hearing loss formulas (0.5, 1, and 2 kHz

Table 8.  Percent of Employees in the ANSI S12–12 Working Group's 24 Audiometric Databases (46,000 Employees) Who Are Potentially Compensable on Their Most Recent Audiograms Using Three Different Hearing Loss Formulas (Low Fence = 25 dB)

| Mean Age, Years | Frequencies Averaged, kHz | | |
|---|---|---|---|
| | 0.5,1,2 | 0.5,1,2,3 | 1,2,3 |
| 37 | 4.7 | 8.7 | 11.5 |

and 0.5, 1, 2, and 3 kHz) are 4.7% and 8.7%, respectively. For the formula utilized by the federal government (1, 2, and 3 kHz) the percentage is 11.5%. It might be expected that for the total U.S. noise-exposed workforce the percentages would be lower than those shown in Table 8, because the ANSI 12–12 databases over-represent the very high noise-exposure industries such as woodworking and textiles. In addition, in populations without any industrial noise exposures, a small percentage of individuals would be expected to exhibit hearing losses that would exceed the low fence for compensation formulas but who would not be able to file for compensation. Our best guess is that the percentage compensable in nonexposed populations with a similar age distribution would be 2.0% or less.

Presented in Figure 21 is the HLc distribution for the three defined hearing loss formulas in one ANSI S12–12 database which exhibits roughly the same percentages of compensable employees as the averages shown in Table 8 for the total ANSI S12–12 population. In Figure 21 the calculated HLc values are represented on the vertical axis, and the horizontal axis shows the cumulative percent of the population with this HLc value or less. For example, using the 0.5, 1, and 2 kHz formula, approximately 72% of this population exhibited a HLc value of 10 dB or less. The horizontal reference line in Figure 21 shows the LOW FENCE (25 dB, 0% disability).

For the population chosen and the 0.5, 1, and 2 kHz formula, 5.2% of employees are potentially compensable (that is, they exhibit HLc values at or above the 25 dB LOW FENCE). As is evident from the figure, a significant percentage of the employees who are potentially compensable have HLc values that fall close to the LOW FENCE of 25 dB.

In studying the data presented in Figure 21, two main observations are evident:

1. The slope of the curve is still sufficiently flat as the curve crosses the LOW FENCE so that factors that affect the measurement of an employee's hearing thresholds can have a significant effect on the percentage of the population that is potentially compensable. Such factors include proper instruction of the employee in how to respond during the audiometric evaluation, proper audiometer calibration, proper utilization of hearing protection devices prior to the

## Cumulative Distribution of Hearing Loss in One Industrial Population

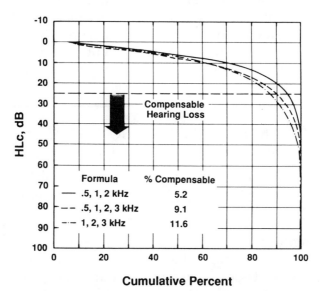

**Figure 21.** Averaged total hearing level (HLc) using three compensation hearing loss formulas versus cumulative percentage of the population with HLc values equal to or less than the selected level.

audiometric evaluation, and the learning effect from repeated audiometric evaluations.

2. Since most employees must anticipate a minimum economic benefit from filing for compensation to make it worth the time and expense involved, in effect there exists an ECONOMIC THRESHOLD FENCE, several dB *higher* than the LOW FENCE, which would have to be exceeded before the typical employee would bother to file. This economic factor significantly reduces the percent of the population that, practically, is really able to file for compensation.

In our experiences with industrial HCPs, we have encountered managers who fear that a flood of hearing loss compensation claims would begin if employees were informed about the amounts of hearing loss shown on their annual audiograms. Considering the data presented in Table 8 and Figure 21, this management attitude is clearly unjustified. For industries where the work force is old (average workforce age around 40 or older) and has a long history of high noise exposure predating the development of HCPs, management does face a significant liability in terms of the potential cost of employee claims for hearing loss due to on-the-job noise exposure. However, for a majority of U.S. industries, the current percentage of the population potentially compensable is most likely very low; less than 3.5%.

> THE POTENTIAL COMPENSATION COSTS DO NOT
> JUSTIFY THE MANAGEMENT PRACTICE OF
> NOT PROVIDING EMPLOYEES WITH
> MEANINGFUL FEEDBACK ABOUT THEIR HEARING.

Meaningful audiometric feedback is the best motivational tool to influence employees to halt the progression of their noise-induced hearing losses. Therefore, it is counter-productive as well as economically unjustified to withhold hearing status information from employees. Frankly, discussing the audiogram results does not lead to a flood of suits. Managers who withhold information are actually more likely to *increase* the probability of compensation claims by fostering suspicion among employees through their secretive attitude.

### EVALUATING WHETHER A HEARING
### LOSS IS WORK-RELATED

In trying to analyze the legitimacy of a compensation claim for occupational hearing loss, we have found that it can be very helpful to use the ISO 1999.2 standard[3] for predicting the hearing of a noise-exposed population. As discussed above, this standard allows the user to predict the resulting hearing levels if a population with a specified susceptibility to noise and to age effects receives a specified noise exposure (the intensity of the noise and the duration in years).

The ISO 1999.2 standard predicts only population hearing level distributions, not hearing thresholds for an individual. However, if the observed hearing thresholds for a claimant do fall *within* the range of population predictions, then the claimant's hearing loss can be accepted as probably work-related. Conversely, if the claimant's hearing loss *exceeds* predictions for a very susceptible percentile of the population, then there is a high probability that nonoccupational factors have contributed to the hearing loss.

To use this standard it is necessary to determine the effective level of noise and the duration of exposure in years. We will not go into the details of how to determine the employee's effective exposure over several years, as these procedures are well defined elsewhere.[3,7] At least one computer program exists [for the Apple and IBM (PC, XT, and AT) or IBM-compatible computers] which makes it easy to use the ISO 1999.2 model to predict hearing thresholds (in audiogram format or for individual frequencies over exposure duration) after a few parameters have been defined and entered. For more information, see Resources listing #9.

As our first example in using ISO 1999.2, consider the hearing levels presented in Figure 22 for a 40-year-old employee who has worked for one company for 20 years. As for many employees whose noise exposures began before OSHA was established, there is no preexposure audiogram available to compare to the current audiogram. However, the employee's current hearing thresholds can be compared to the range of hearing thresholds predicted for different susceptibility percentiles of a population with the same noise exposure as the claimant.

Let's assume that the employee's effective noise exposure was 85 dB(A) over the 20 years. Shown in Figure 22 are the predicted hearing levels for this exposure for three population percentiles: the 95th percentile (95% of the population would exhibit worse hearing ability), 50th percentile (50% would exhibit worse hearing), and the 5th percentile (only 5% would exhibit worse hearing). Since the employee shows hearing thresholds far poorer than predicted for the 5th percentile, we would conclude that the employee's present hearing loss did not result solely from employment at this company.

What if the employee's effective noise exposure had been higher—100 dB(A) for 20 years? Consider the data presented in Figure 23, where the same hearing thresholds for the employee are compared to population percentile predictions for this higher

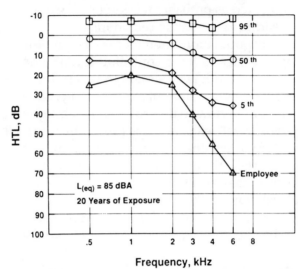

**Figure 22.** Measured hearing thresholds for one employee versus predicted hearing levels for population precentiles of low, average, and high susceptibility (95th, 50th, and 5th percentiles) exposed to an $L_{eq}$ of 85 dB(A) for 20 years.

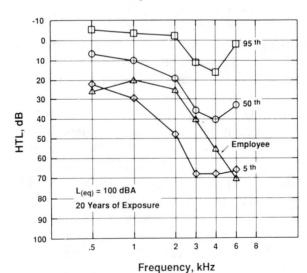

**Figure 23.** Measured hearing thresholds for one employee versus predicted hearing levels for population percentiles of low, average, and high susceptibility (95th, 50th, and 5th percentiles) exposed to an $L_{eq}$ of 100 dB(A) for 20 years.

exposure. In this case the employee's hearing levels do lie within the range predicted using ISO 1999.2. Therefore, we would conclude that the employee's on-the-job noise exposure probably did cause the hearing loss (assuming of course that causation by medical pathology or nonoccupational noise had been ruled out).

Another way to use the ISO 1999.2 model is to predict change in hearing over a period of years. To illustrate this approach, we will now discuss the case of a male employee who had relatively good hearing at age 24, but who exhibited a very significant decline in his hearing over the next couple of years.

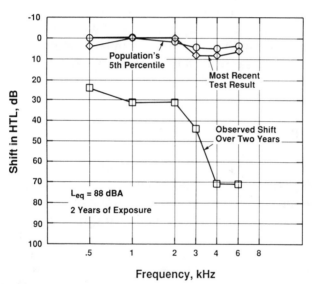

**Figure 24.** Actual shift in hearing thresholds for one employee versus predicted hearing shift for susceptible ears (5th percentile) exposed to an $L_{eq}$ of 88 dB(A) for two years.

This employee worked in a production area where the effective daily TWA was 88 dB(A) or less. Shown in Figure 24 are the shifts in hearing levels predicted for the 5th percentile of the population exposed for 2 years at 88 dB(A), and the dramatic shifts in hearing actually shown by the employee over this two-year period. We would conclude that the shifts observed in the employee's hearing were not due to on-the-job noise exposures.

This employee was appropriately flagged by the company's HCP and was sent at the company's expense for evaluation by an otologist (ear specialist). No medical cause for the large hearing change could be established, and no treatment was recommended. However, on the next HCP audiogram, his hearing returned to within 10 dB of his original thresholds, and it has remained stable over the most recent three years. The remaining small shifts from the original thresholds are shown in Figure 24 by the line labeled "most recent test."

When questioned about this unusual pattern of hearing decline and subsequent recovery, the employee reported that he had been taking diet pills called "Dynamic Fat Burners" when the loss occurred. When he stopped taking this over-the-counter medication, his hearing started returning to normal.

Although this example is a rare case, it illustrates how the ISO 1999.2 model can be used in a similar manner to rule out occupational causation of a hearing change due to gunfire or to nonoccupational acoustic trauma.

## ADBA PROVIDES THE PREVENTIVE MEDICINE

The most effective way to prevent employees from developing potentially compensable hearing loss from on-the-job noise is to utilize the audiometric database to monitor HCP effectiveness and warn management when more protective HCP policies are required. Recall the discussion in Chapter 8 of the benefits of utilizing audiometric database analysis (ADBA) as a preventive measure. Turn back to Chapter 8 to review Figure 17, which illustrates how ADBA results can identify HCP problems before employees develop significant hearing loss.

Also review Figure 19 (earlier in this chapter) to recall how the NIPTS component of total hearing loss develops rapidly early during the worker's noise exposure history. If the employee is young and has no significant age-related hearing loss yet, developing NIPTS may not seem to pose a problem from the standpoint of potential compensation liability. However, keep in mind that the AGE EFFECT component, as shown in Figure 18, will begin to grow more rapidly during the later stages of the employee's career and will push the total hearing loss over the LOW FENCE. We cannot control or prevent AGE EFFECTS, but we *can* prevent NIPTS. Individual audiogram review will eventually identify those susceptible individuals who show significant NIPTS, but ADBA can identify an inadequate HCP *before* a large segment of the population develops significant NIPTS that may not be detected quickly by individual audiogram review. Smart management will use ADBA to zero in on HCP problems early, then implement corrective actions to prevent the company from developing compensation liability.

## ADBA PROVIDES SUPPORTIVE OR NONSUPPORTIVE LEGAL DATA

As of the date of the publication of this book, the authors are not aware of any instance in which management has used the information in the audiometric database in its defense against an employee's claim for compensation for on-the-job noise-induced hearing loss. We are aware of a few HCPs where management tried very hard to achieve program effectiveness, and ADBA analysis supported their efforts. If an employee in one of these plants were to develop hearing loss due to off-the-job noise exposure or in work areas where the daily TWAs are very low, then management could present the ADBA findings to support the fact that an effective HCP was in place during the time that the compensable hearing loss supposedly occurred. (Of course, personal identification numbers for the employees would be removed from the database for privacy purposes before the findings of ADBA were utilized as a defense.)

It is also possible to use ADBA results to demonstrate that an HCP has *not* been effective in protecting employees. We believe that this approach will become common in the future. That is, when an employee develops a compensable hearing loss, the lawyers for the plaintiff will insist that the company submit the audiometric database for evaluation by an impartial source. If company management claims that an effective HCP has been implemented, this defense would be invalidated if the ADBA results indicate otherwise. When this approach is accepted by the judicial process, management will suddenly renew its interest in the mechanics and effectiveness of the company's HCP.

Management must realize that contained in its audiometric database is the history of its HCP. This history information can provide evidence as to the validity of a claim for compensation for an occupational hearing loss.

> THE AUDIOMETRIC RECORDS ARE A
>
> HISTORY OF THE COMPANY'S
>
> SUCCESSES AND/OR FAILURES

## THE AUDIOMETRIC RECORDS TELL IT LIKE IT IS

There should be little doubt at this point about the essential need for proper audiometric testing procedures and a periodic evaluation of their validity. To end up in court with inadequate employee audiograms is almost an admission of guilt. As an example, the audiometric history of Mr. Reece Clark is presented in Figure 25. Mr. Clark had filed for compensation for on-the-job noise-induced hearing loss, and his audiometric records were entered as evidence.

Obviously, Mr. Clark's audiograms are erratic and unacceptable in quality up until at least the fifth evaluation. It is evident that up to the fifth test the responsible persons for this HCP were not doing an adequate job—otherwise they would have detected problems with Mr. Clark's test results and re-tested him until more consistent findings were obtained.

There are significant consequences of the failure to catch and correct the audiometric testing related problems indicated in Figure 25. First, the company finds it very difficult to argue that its HCP was effective back during the time when Mr. Clark's hearing loss was developing. Second, the data presented lend support to claims made by other employees of the company, who also may file for compensation for noise-induced hearing loss, that the HCP was ineffective during this time period. As was pointed out earlier, the audiometric database is a history of what the company has or has not done with respect to properly protecting the employees from on-the-job noise-induced hearing loss.

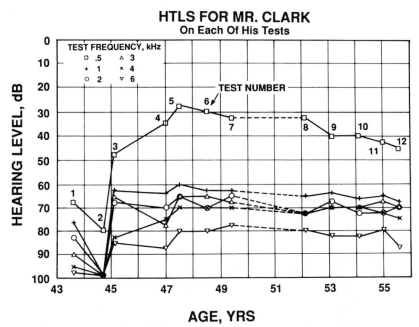

**Figure 25.** Hearing thresholds measured for Mr. Clark on each of his audiograms, plotted as a function of age.

Also note in Figure 25 the apparent calibration shift that occurred between test numbers 4 and 5. When the hearing levels of a population or individual shift strongly in the same direction between approximately annual evaluations, as opposed to a gradual change due to the learning effect, normal aging, hearing loss or a combination thereof, then a change in the calibration reference should be one suspected source.

The result is that this company, if Mr. Clark's audiometric record is typical, finally got its act together around the time that Mr. Clark received his fifth audiometric evaluation. As a consequence, the company does not have good records on what actually happened to Mr. Clark's hearing prior to the time of his fifth audiogram. Therefore, it does not have any valid evidence, in our opinion, as to how Mr. Clark's hearing ability had changed from the time of his first audiometric evaluation. Again, the audiometric record is a history of the company's mistakes and/or accomplishments.

## THE ESSENTIAL AUDIOGRAMS AND DOCUMENTATION

What audiograms are necessary in order to establish an adequate defense against an unjustified claim for on-the-job noise-induced hearing loss? The answer is simple: all audiograms that are necessary to prove that adequate HCP procedures were implemented, and to document whether the employee's hearing loss occurred while employed at your company. Therefore, the following audiogram records should be maintained as long as the employee can legally file a claim:

1. Preemployment audiograms—Document that you gave the employee a copy and also sent a copy to the previous employer if a compensable hearing loss is identified. It is important that you document and explain to the employee before hiring that a potential compensable hearing loss exists, and that his previous employer(s) may be responsible for his present hearing loss. If a potentially compensable loss is found at the time of hire, any available past audiograms should be obtained from military service and/or previous employers.

2. Regular monitoring audiograms—Annual audiograms are adequate for most noise exposures. Semiannual audiograms are recommended during the first two years of exposure for employees with TWAs of 100 dB(A) or higher, in order to identify susceptible individuals who are inadequately protected before significant hearing loss develops. We recommend this because NIPTS can develop so quickly in susceptible ears with high noise exposures, as shown in Figure 19.

3. Auditory history and medical records—At the time of hire the employee should be asked to describe previous occupational and nonoccupational noise exposures, as well as medical conditions related to hearing. This auditory history information should be updated annually at audiogram time. If the employee is referred for otological evaluation or for a clinical audiogram, the resulting reports should be obtained and kept.

4. Rehire audiograms—When the employee takes a leave of absence or is temporarily laid off, an audiogram is needed upon rehire to reestablish the hearing thresholds, since the individual could have sustained a hearing loss while off your payroll. Without adequate documentation of changes that occurred, your company will be responsible.

5. Termination audiograms—If the employee leaves the company he/she should not be paid until a satisfactory termination audiogram is obtained.

6. Past Records—If an employee notifies the company that he/she intends to file, then any known previous audiometric findings should be obtained.

Note that the records needed for compensation purposes exceed the recordkeeping requirements of the OSHA regulations. (Refer back to Table 4 in Chapter 8 for additional information.) The two systems—OSHA and Workers' Compensation—are separate and unrelated. The employer can administer an HCP that meets minimum OSHA compliance requirements, yet may be ineffective in preventing hearing loss and result in company compensation liability. Only an effective HCP can reduce liability.

## HEARING PROTECTION—THE WEAKEST DEFENSE LINK

Employees who file for compensation for occupational hearing loss today should have been wearing hearing protection devices for many years, though probably not from the beginning of their work careers. In some compensation hearings, management has attempted to argue that the employee could not have sustained a hearing loss on-the-job since HPDs were provided, and, according to the NRR published by the manufacturer, the HPDs should have reduced the employee's exposure to nonhazardous levels. These arguments have failed for two important reasons.

First, the protection an employee achieves using HPDs in the real world is very much less than indicated by the NRR (refer back to Chapter 6). Second, the employee may receive little or no protection from HPDs unless they are properly fitted, replaced on a regular basis, and the user is both trained in the correct way to use them and supervised in using them consistently. Management cannot claim that the employee received any significant degree of protection unless the company can document the existence of an effective HCP in which all these functions were carried out. For management to approach a compensation case claiming that the level of protection provided by HPDs is equal to the NRR is almost guaranteed to diminish, if not destroy, the company's claim of having an effective HCP. It also vividly points out management's lack of knowledge of the available literature in the hearing conservation area. The claimant's attorney will simply put an expert on the stand and have this individual review the extensive world literature that demonstrates the limitations of HPDs as utilized in real-world environments.

The best defense against criticism of hearing protection is ADBA results which verify that the company has an effective HCP. ADBA has been utilized to rank the relative effectiveness of different HPDs and to show that HPDs can prevent hearing loss if properly fitted and issued, and if correct and consistent utilization is enforced.

In order to build a supportable defense, management should claim only 50% of the manufacturers' published NRRs as the expected level of protection achieved in real-world environments.


## APPROPRIATELY EDUCATE YOUR TEAM

Most individuals who play a major role in the company's HCP will have built up a storehouse of beliefs about hearing conservation; however, in some instances the "information" learned is erroneous. In more than one instance, we have been told that the company was using a particular HPD because it was a universal fit device which did not need to be fitted. We have heard this claim made not only for earmuffs and foam earplugs, but also for medium-size triple-flange plugs. Unfortunately, there is *no* universal-fit HPD, and industrial personnel who believe otherwise are failing to protect their employees. Therefore, all company personnel involved in the HCP need training in their HCP tasks in order to carry out their responsibilities correctly.

Likewise, in preparation of the defense against a hearing loss compensation claim, the company should make sure that the lawyer(s) involved receive appropriate training about hearing conservation prior to preparing for the court hearings. As anyone who has had the unfortunate opportunity to become involved in court proceedings knows, the facts often get twisted, and the outcome may depend upon nontechnical factors, including the ability of the lawyer to handle HCP-related terminology. If the lawyer is not properly trained, then the proper questions are not asked, appropriate objections are not made, and the lawyer's expert witness is not able to present important facts to the judge and/or jury.

It is critical, especially in the case of a trial by jury, that the expert witnesses be able to translate the often highly technical terms into wording that the jury members (average people on the street) can understand.

Prior to the time of depositions or trial, the lawyer should spend at least two days with the expert(s) to familiarize her/himself with HCP terminology, particularly the terms related to the effects of noise on hearing.

Other important experts, who may testify or give written statements rather than appearing in court, are the audiologist and/or physician who examine the employee to evaluate the degree of hearing loss and its possible causes. Our experience has clearly shown that even otologists are limited in their understanding of how noise-induced hearing loss develops and of the cause-effect relationship between noise exposure and the degree of loss. If a company has been sending its employees to the local MD or audiologist over the past 20 years, and these individuals typically reference the on-the-job noise exposures as the only explanation for all observed hearing losses, then the comments made in the employees' records are in fact an admission that the HCP has not been effective. It is important that these professionals develop adequate knowledge of the areas of hearing conservation and the effects of noise. Otherwise the company pays at least twice: first when the employee is referred, and again when the employee files for compensation.

> WITH INADEQUATELY PREPARED PROFESSIONALS,
>
> THE EMPLOYER PAYS TWICE:
>
> ONCE WHEN THE EMPLOYEE IS REFERRED,
>
> AND AGAIN WHEN HE FILES FOR COMPENSATION.

## SHOULD MANAGEMENT ALLOW AN EMPLOYEE TO DEVELOP A COMPENSABLE NOISE-INDUCED HEARING LOSS?

The answers are NO, YES, and MAYBE!

NO! If an employee joins your work force with good hearing, then the company's HCP practices should, in general, prevent this worker from developing a compensable hearing loss due to on-the-job noise exposure. If the company prevents occupational hearing loss in its population, then there is only a very small chance that the worker will develop a potentially compensable hearing loss, as defined using the 0.5, 1, 2, and 3 kHz formula, due to off-the-job exposures while working at your company.

YES! If a person applies for work at your facility who is a desirable employee, but who also exhibits a hearing loss that is compensable under existing regulations, you should still hire this individual. Document the preemployment hearing thresholds in a valid audiogram. Although the employee's hearing will get worse over time due to the AGE EFFECT component of hearing loss, in most jurisdictions your company will be legally liable only for the *increase* in hearing loss, not for the preexisting loss. However, the services of an experienced employee are well worth the risk that the worker may one day file for compensation. Good workers are simply worth this additional small potential cost of doing business.

MAYBE! This one is on you . . .

# REFERENCES

1. Cudworth, A. L. "Workers' Compensation," in *Noise & Hearing Conservation Manual*, 4th ed., E. H. Berger, W. D. Ward, J. C. Morrill, and L. H. Royster, Eds. (Akron, OH: American Industrial Hygiene Association, 1986).
2. Suter, A. H. "Calculation of Impairment or Handicap," in *Forensic Audiology*, M. B. Kramer and J. M. Armbruster, Eds. (Baltimore, MD: University Park Press, 1982).
3. International Standardization Organization. ISO 1999.2: "Acoustics—Determination of Noise Exposure and Estimation of Noise-Induced Hearing Impariment," (Geneva, Switzerland: ISO, 1989).
4. Royster, L. H., and J. D. Royster. "Making the Most Out of the Audiometric Data Base," *Sound and Vibration* 18(5) 18–24 (1984).
5. Royster, L. H., and J. D. Royster. "Getting Started in Audiometric Data Base Analysis," *Seminars in Hearing* 9:325–338 (1988).
6. Royster, J. D., and L. H. Royster. "Audiometric Data Base Analysis," in *Noise & Hearing Conservation Manual*, 4th ed., E. H. Berger, W. D. Ward, J. C. Morrill, and L. H. Royster, Eds. (Akron, OH: American Industrial Hygiene Association, 1986).
7. Royster, L. H., E. H. Berger, and J. D. Royster. "Noise Surveys and Data Analysis," in *Noise & Hearing Conservation Manual*, 4th ed., E. H. Berger, W. D. Ward, J. C. Morrill, and L. H. Royster, Eds. (Akron, OH: American Industrial Hygiene Association, 1986).

# Resources For Industrial Hearing Conservation

1. American National Standards Institute (ANSI)
   1430 Broadway, New York, NY 10018
   (212) 354-3300

2. Council for Accreditation in Occupational Hearing
   Conservation (CAOHC)
   66 Morris Avenue, Springfield, NJ 07081
   (201) 379-1100

3. Sound and Vibration: the Noise and Vibration Control
   Magazine
   P.O. Box 40416, Bay Village, OH 44140
   (216) 835-0101
   [Subscription free of charge to qualified individuals]

4. E-A-R Division, Cabot Corporation
   Box 68898
   Indianapolis, IN 46268-0898
   (317) 872-1111
   [Series of 19 EARlogs (reference pamphlets concerning
   practical topics in hearing protection and hearing
   conservation) available on request]

5. ANSI S12–12 Working Group for Evaluation of Hearing
   Conservation Programs
   Contact Larry Royster, Working Group Chairman
   Dept. of Mechanical & Aerospace Engineering
   North Carolina State University
   P.O. Box 7910
   Raleigh, NC 27695-7910
   (919) 737-3024

6. National Hearing Conservation Association (NHCA)
   900 Des Moines Street, Suite 200
   Des Moines, IA 50309
   (515) 266-2189

7. American Speech-Language-Hearing Association (ASHA)
   10801 Rockville Pike
   Rockville, MD 20852
   (301) 897-5700

8. American Industrial Hygiene Association (AIHA)
   475 Wolf Ledges Parkway
   Akron, OH 44311-1087
   (216) 762-7294

9. The N.A.N. Computer Program for Estimating a Population's
   Hearing Threshold Characteristics (version 2.0)
   Environmental Noise Consultants, Inc.
   P.O. Box 144
   Cary, NC 27512-0144
   (919) 782-1624

# Index